LIST OF TITLES

Already published

Cell Differentiation	J.M. Ashworth
Biochemical Genetics	R.A. Woods
Functions of Biological Membranes	M. Davies
Cellular Development	D. Garrod
Brain Biochemistry	H.S. Bachelard
Immunochemistry	M.W. Steward
The Selectivity of Drugs	A. Albert
Biomechanics	R. McN. Alexander
Molecular Virology	T.H. Pennington, D.A. Ritchie
Hormone Action	A. Malkinson
Cellular Recognition	M.F. Greaves
Cytogenetics of Man and other Animals	A. McDermott
RNA Biosynthesis	R.H. Burdon
Protein Biosynthesis	A.E. Smith
Biological Energy Conservation	C. Jones
Control of Enzyme Activity	P. Cohen
Metabolic Regulation	R. Denton, C.I. Pogson
Plant Cytogenetics	D.M. Moore
Population Genetics	L.M. Cook
Insect Biochemistry	H.H. Rees
A Biochemical Approach to Nutrition	R.A. Freedland, S. Briggs
Enzyme Kinetics	P.C. Engel
Polysaccharide Shapes	D.A. Rees
Transport Phenomena in Plants	D.A. Baker
Cellular Degradative Processes	R.T. Dean
Human Genetics	J.H. Edwards
Human Evolution	B.A. Wood
Metals in Biochemistry	P.M. Harrison, R. Hoare
Isoenzymes in Biology	C.C. Rider, C.B. Taylor
Genetic Engineering: Cloning DNA	D. Glover

In preparation

The Cell Cycle	S. Shall
Bacterial Taxonomy	D. Jones, M. Goodfellow
Biochemical Systematics	J.B. Harbourne
Membrane Assembly	J. Haslam
Invertebrate Nervous Systems	G. Lunt

Editors' Foreword

The student of biological science in his final years as an undergraduate and his first years as a graduate is expected to gain some familiarity with current research at the frontiers of his discipline. New research work is published in a perplexing diversity of publications and is inevitably concerned with the minutiae of the subject. The sheer number of research journals and papers also causes confusion and difficulties of assimilation. Review articles usually presuppose a background knowledge of the field and are inevitably rather restricted in scope. There is thus a need for short but authoritative introductions to those areas of modern biological research which are either not dealt with in standard introductory textbooks or are not dealt with in sufficient detail to enable the student to go on from them to read scholarly reviews with profit. This series of books is designed to satisfy this need. The authors have been asked to produce a brief outline of their subject assuming that their readers will have read and remembered much of a standard introductory textbook of biology. This outline then sets out to provide by building on this basis, the conceptual framework within which modern research work is progressing and aims to give the reader an indication of the problems, both conceptual and practical, which must be overcome if progress is to be maintained. We hope that students will go on to read the more detailed reviews and articles to which reference is made with a greater insight and understanding of how they fit into the overall scheme of modern research effort and may thus be helped to choose where to make their own contribution to this effort. These books are guidebooks, not textbooks. Modern research pays scant regard for the academic divisions into which biological teaching and introductory textbooks must, to a certain extent, be divided. We have thus concentrated in this series on providing guides to those areas which fall between, or which involve, several different academic disciplines. It is here that the gap between the textbook and the research paper is widest and where the need for guidance is greatest. In so doing we hope to have extended or supplemented but not supplanted main texts, and to have given students assistance in seeing how modern biological research is progressing, while at the same time providing a foundation for self help in the achievement of successful examination results.

General Editors:

W. J. Brammar, Professor of Biochemistry, University of Leicester, UK

M. Edidin, Professor of Biology, Johns Hopkins University, Baltimore, USA

Genetic Engineering Cloning DNA

David M. Glover

Cancer Research Campaign Eukaryotic Molecular Genetics Group,
Department of Biochemistry,
Imperial College of Science and Technology.
London

1980

London and New York

Chapman and Hall

150th Anniversary

First published 1980
by Chapman and Hall Ltd.
11 New Fetter Lane, London, EC4P 4EE
Published in the U.S.A. by
Chapman and Hall Ltd
in association with
Methuen, Inc.,
733 Third Avenue, New York, NY 10017
© *1980 D. M. Glover*

Printed in Great Britain at the
University Printing House, Cambridge

ISBN 0 412 16170 2

British Library Cataloguing in Publication Data

Glover, David M
    Genetic engineering. – (Outline studies in biology).
    1. Molecular cloning   2. Recombinant DNA
    I. Title   II. Series
    574.8'732   QH442.2   80-40660

    ISBN 0-412-16170-2

Contents

Abbreviations

Ap^r	Ampicillin resistant
Ap^s	Ampicillin sensitive
Cm^r	Chloramphenicol resistant
Cm^s	Chloramphenicol sensitive
Kb	Kilobases – 1000 bases or base pairs of single or double stranded nucleic acids respectively.
P_L	Leftward promoter of phage λ
P_R	Rightward promoter of phage λ
P_R'	Late rightward promoter of phage λ
Pre	Phage λ promoter for the establishment of lysogeny
Prm	Phage λ promoter for the maintenance of lysogeny
SV40	Simian Virus 40
Tc^r	Tetracycline resistant
Tc^s	Tetracycline sensitive

1   Introductory remarks

Past progress in understanding the molecular biology of prokaryotic gene expression has relied heavily upon studies involving bacteriophage and bacterial plasmids. Of the bacteriophage themselves, the *E. coli* phage λ is perhaps the best characterised. The interaction of phage λ with the host cell is a particularly fruitful area of study, as here are a set of genes which can either direct cell lysis or become stably associated with the host chromosome in lysogeny. In the production of infectious phage from lysogens, the excision of the phage λ genome from the *E. coli* chromosome is usually precise. Occasionally, however, the excision is imperfect and results in a λ phage transductant which carries that segment of the bacterial chromosome which was adjacent to the phage attachment site. Such specialised transducing phage have been invaluable, providing the means to assay for specific messenger RNAs by nucleic acid hybridisation or enabling the production of large amounts of particular gene products. Research on the bacterial plasmids has had a similar history. The discovery and the rationalisation of the mechanism whereby F factors promote bacterial conjugation was central to the development of *E. coli* genetics. Just as the imperfect excisions of phage λ from its lysogenic state can result in a circular phage genome carrying a segment of bacterial DNA, so the imperfect excision of an F plasmid from an Hfr strain results in an F′ plasmid which also carries a segment of bacterial DNA. Such F′ plasmids have been invaluable 'vectors' for carrying specific genes from one *E. coli* strain to another and have perhaps been most useful in the construction of merodiploid strains which have enabled the elucidation of the control circuits of many bacterial operons.

  The principles of genetic engineering which are described in this book are analogous to these 'natural' events, but they overcome the limitation of an absolute dependence upon the *in vivo* recombinational mechanisms of the *E. coli* cell. The techniques for recombining DNA *in vitro* make it possible to insert DNA from any organism into a plasmid or viral replicon to form a chimaeric molecule which can replicate in the host organism for that replicon, be it a prokaryote or a eukaryote. In most cloning experiments a heterogeneous population of *in vitro* recombinant DNA molecules is first generated. When, for example, an *E. coli* plasmid vector is used the recombinant molecules must retain two properties of the plasmid: an ability to replicate autonomously, and a marker function which will allow the selection of cells transformed by the plasmid. The conditions of such a transformation are adjusted so that an individual bacterial cell only receives a single plasmid molecule. A homogeneous population of recombinant plasmid DNA molecules can then be prepared from a culture derived from a single transformed colony. In the case of a viral vector, such as bacteriophage λ, homogeneous isolates of the segments of foreign DNA are obtained from phage picked from individual plaques produced by single recombinant bacteriophage DNA molecules.

Most studies carried out to date with cloned DNA segments have exploited the cloning technology as a means of preparing large quantities of specific DNA sequences from complex genomes. Detailed physical maps of these specific genes can then be constructed, and such studies have begun to give us some understanding of eukaryotic genomes. One avenue of research which remains to be fully exploited is the use of cloned DNAs to test functional relationships of DNA sequences, and so investigate the mechanisms controlling the expression of eukaryotic genes. A major application of the *in vitro* recombinant DNA technology which has excited the imagination of industrialists is the potential to design microorganisms which could produce polypeptides of industrial or pharmacological importance.

These potential benefits were rapidly recognised, as also were the potential hazards. Some of the first *in vitro* recombinants consisted of segments of *E. coli* DNA carrying the galactose operon linked to the DNA of the mammalian tumour virus SV40 [1]. These molecules were never introduced into *E. coli* because of the hypothetical hazard of propagating tumour virus DNA in bacteria which survive in the intestinal tract. Early assessments of some of these problems are found in the Ashby Committee report [2] and in the summary statement of the Asilomar Conference on recombinant DNA molecules [3]. In the absence of evidence viewpoints became sharply polarised, and the whole issue provoked much debate and many reports from both governmental and scientific bodies [4, 5, 6]. The main fear was that a potentially hazardous gene from a eukaryote might inadvertently be cloned. The *E. coli* host cells might then successfully colonise the intestinal tracts of animals and precipitate some disastrous pandemic. Another point of view was that prokaryotic organisms within nature were frequently in contact with, and must take up, eukaryotic DNA from decaying plant and animal matter. It is then likely, given the enormity of the earth's population of prokaryotes, that recombination of the type that can now be carried out *in vitro* has already had chance to occur, and that the recombinant organisms have no selective advantage. Largely as a result of the exploitation of *in vitro* recombination techniques, we now know that most eukaryotic genes have a pattern of chromosomal organisation that precludes their expression in prokaryotic cells (see Chapter 6). In order to get expression of eukaryotic genes in prokaryotes it is usually necessary to clone either a DNA complement of the mRNA of interest or a chemically synthesised gene correctly linked to prokaryotic signals for the promotion of transcription and the initiation of translation (Chapter 5). The concern about the biological safety of these experiments has, furthermore, stimulated work to develop 'safe' host-vector systems. As a result of such work it has emerged that the standard strain of *E. coli*, after several decades of culture as a laboratory organism, has now a very low probability of survival within the human gut. This probability is reduced even further in the biologically 'safe' host-vector systems which have been developed and will be described in Chapters 3 and 4. As more and more experience has been accrued in this field over the last few years, and as our knowledge of gene organisation and the control of gene expression improves, a more realistic outlook on the hazards of these experiments has developed. The experiments are now carried out in specially designed laboratories under the recommendations of governmental agencies. National and local safety committees are now well established in most countries to monitor the work and to advise on aspects of laboratory safety.

References

[1] Jackson, D. A., Symons, R. M. and Berg, P. (1972), *Proc. natn Acad. Sci. USA.* **69**, 2904.
[2] Ashby, Lord (1975), *Cmnd. 5880, H. M. S. O.* London.
[3] Berg, P. Baltimore, D. Brenner, S., Roblin, R. O. and Singer, M. F. (1975), *Science* **192**, 938.
[4] Williams, R. E. O. (1976), *Cmnd. 6600, H. M. S. O.* London.
[5] Revised N. I. H. Guidelines. *United States Federal Register* **44**, 69210.
[6] *Nature* (1978), **276**, 104.

2 The enzymology of *in vitro* DNA recombination

2.1 Restriction endonucleases

The ease with which DNA molecules can now be joined *in vitro* is a consequence of the availability of restriction endonucleases, enzymes which recognise specific sequences in DNA and then cleave both strands of the duplex. These enzymes have been found in many prokaryotes and are likely to be responsible for the degradation of 'alien' DNA molecules, the indigenous DNA being protected from degradation by a modification enzyme, usually a methylase. Restriction endonucleases are responsible for the phenomenon of host controlled modification of bacteriophage, first described in the early 50s [reviewed in Ref. 1]. If phage λ, which has been propagated on *E. coli* strain K, is then allowed to infect *E. coli* strain B, the efficiency of the infectious process is very low. The phage produced from this infection can, however, reinfect *E. coli* strain B with high efficiency. Three genetic loci can be identified which control this system: *hsd*S, *hsd*M and *hsd*R. A polypeptide which governs the specificity of the system is determined by *hsd*S. The gene product of *hsd*M is the modification enzyme which also interacts with the product of the *hsd*R gene, the restriction endonuclease, in the cleavage process. In the above examples the phage grown on strain K would have been modified at sites specified by the K restriction-modification system. In the first infective cycle in *E. coli* B cells the B restriction-modification system detects the absence of B modification and degrades the infecting DNA. A small proportion of molecules are, however, methylated by the B modification system, and these survive restriction on the next infective cycle. This phenomenon has to be borne in mind when introducing foreign unmodified *in vitro* recombinant DNA into *E. coli*. In order for these molecules to survive the recipient strain should have defective *hsd*S or *hsd*R genes.

The B–K restriction-modification enzyme systems of *E. coli* have been termed class I enzymes: they require Mg^{++}, S-adenosylmethionine and ATP as cofactors, and although they recognise specific sites within the DNA they do not cleave at these sites [2]. A second class of restriction endonucleases has been identified which have simple cofactor requirements and cleave DNA at, or near, specific sequences that are usually several nucleotides long and rotationally symmetrical about the central nucleotide pairs. These latter enzymes have been isolated from a wide range of prokaryotic microorganisms and are invaluable for cloning DNA.

An extensive list of these enzymes, and the sequences recognised by them, has been published by Roberts [3]. Let us only consider the recognition sites of some of the more commonly used restriction endonucleases. In general, these enzymes cleave DNA to generate a nick with a 5′ phosphoryl and 3′ hydroxyl terminus. In some cases the cleavages in the two strands are staggered, and because of the symmetry of the recognition sequence this generates mutually cohesive termini. The plasmid encoded *E. coli* enzyme *Eco*RI is an example of such an enzyme, and cleaves the sequence GAATTC between the G and A residues (see also Fig. 2.1). In the case of *Eco*RI the protruding single stranded ends have 5′ termini. Other enzymes such as *Pst*I, isolated from *Providencia stuartii*, have a staggered cleavage that generates single stranded 3′ termini. There are also enzymes such as *Hae*III from *Haemophilus aegypticus* which generate flush ends. The type II restriction endonucleases have provided us with the means of dissecting simple genomes or cloned segments from complex genomes. I will review such applications of restriction endonucleases in Chapter 6, and concentrate in this chapter upon the uses of these enzymes in processes of *in vitro* DNA recombination.

2.2 Joining restriction fragments with DNA ligase
The most widely used procedure for recombining DNA *in vitro* makes use of those restriction enzymes which generate mutually cohesive termini on DNA. This type of scission was first recognised by Mertz

Table 2.1 Commonly used restriction endonucleases which generate cohesive termini

Enzyme	Microorganism	Cleavage site
BamHI	Bacillus amyloliquefaciens H	G↓GATCC
BglII	Bacillus globigii	A↓GATCT
EcoRI	Escherichia coli RY13	G↓AATTC
HindIII	Haemophilus influenzae Rd	A↓AGCTT
MboI	Moraxella bovis	↓GATC
PstI	Providencia stuartii 164	CTGCA↓G
SalI	Streptomyces albus G	G↓TCGAC

and Davis [4] who showed by electron microscopy that EcoRI-cleaved DNA would cyclise at low temperature. Furthermore they were able to show that the cohesive termini could be covalently sealed with E. coli DNA ligase, and were able to construct recombinant DNA molecules of the bacterial plasmid λdvgal and DNA of SV40.

A number of enzymes are now known to produce cohesive termini (Table 2.1). Bacteriophage or plasmid vectors have been described which will permit the cloning of fragments generated by many of these enzymes. Some of these enzymes share common central tetranucleotides in their recognition sequence e.g. BamHI, BglII and MboI, and so although these enzymes recognise different sites in DNA, they all produce identical single stranded 5' tails which allow the joining of fragments generated by the different enzymes within this set. The identical nature of the termini of DNA fragments from any organism following restriction endonuclease cleavage is the very property which permits the annealing and subsequent ligation of DNA from diverse sources. The general principle of this cloning approach is illustrated in Fig. 2.1 for the specific case of cloning EcoRI fragments of Drosophila DNA in the bacterial plasmid pSC101. In certain experiments the indiscriminate joining which this allows can be disadvantageous. This would be the case if one, for example, wished to reconstruct a DNA molecule in which the restriction fragments were in a particular order so as to

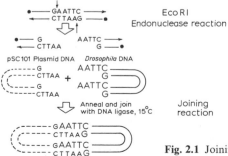

Fig. 2.1 Joining EcoRI fragments with DNA ligase

11

regenerate a transcription unit. One potential way of overcoming this problem has not yet found widespread application. This is to anneal DNA which has been cleaved with an enzyme such as *Hga*I. This enzyme recognises the sequence GACGC, but rather than cleaving the DNA at this site it makes staggered single strand scissions 5 and 10 nucleotides away, respectively, from the recognition site. This generates a set of fragments which have unique cohesive ends and which can therefore only be reassembled in one particular order.

DNA ligase has the physiological role of sealing single strand nicks in DNA which have 5' phosphoryl and 3' hydroxyl termini and which are generated both as a result of the discontinuity of the replication fork and also in repair processes. The two enzymes which are extensively used for covalently joining restriction fragments are the ligase from *E. coli* and the enzyme encoded by phage T4. The *E. coli* enzyme uses NAD as cofactor whereas the T4 enzyme uses ATP, but in either case the cofactor serves to adenylate the ε-NH$_2$ of a lysine residue in the enzyme. The 5' phosphoryl terminus of the DNA is then adenylated by the enzyme–cofactor complex and finally a phosphodiester bond is formed with the liberation of AMP (Fig. 2.2) [5]. The enzyme purified from T4 infected *E. coli* has been used most extensively since it is easier to prepare and, unlike the *E. coli* enzyme, it is available commercially. The T4 enzyme has the additional advantage that, with high concentrations of enzyme and of ATP, it will join DNA molecules cleaved by restriction enzymes which generate fully base paired 'flush ends'. In this case the molecules to be joined are not held together by hydrogen bonds between mutually cohesive termini [6]. The ease by which this enzyme can be prepared has been facilitated by the cloning of its gene in phage λ vectors (see Chapter 5).

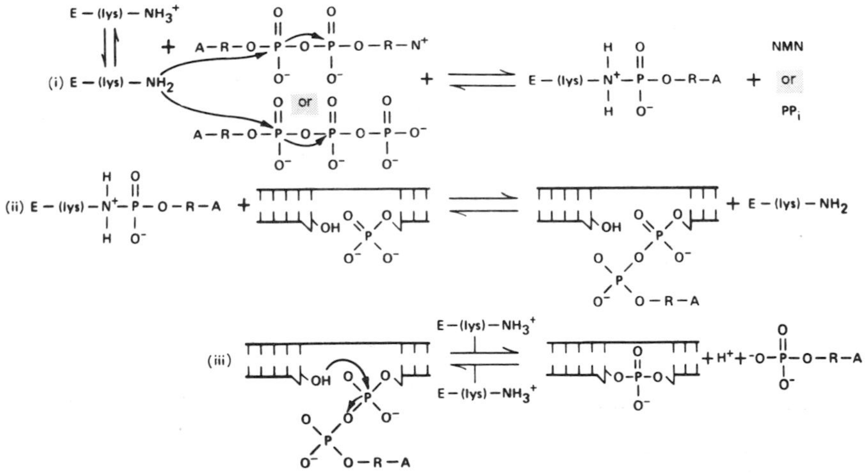

Fig. 2.2 Mechanism of DNA ligase reaction (from [5]; copyright 1974 American Association for the Advancement of Science).

One major disadvantage of joining a plasmid vector to foreign DNA at cohesive ends generated by restriction endonucleases is the frequency of self-cyclisation of the vector plasmid. This results in a 'background' of transformed colonies which contain only the vector plasmid. This can be overcome by treating the restricted plasmid with either bacterial or calf intestinal alkaline phosphatase in order to remove the terminal 5′ phosphoryl groups. The two ends of the plasmid vector will then be unable to be covalently joined by DNA ligase. The restriction fragments of the foreign DNA are not, however, treated with phosphatase and so their 5′ phosphoryl groups can be covalently joined to the 3′ hydroxyl groups of the plasmid. This results in hybrid molecules in which, at each site of ligation, the vector is joined to the foreign DNA in one strand only whilst the other strand has a nick with 3′ and 5′ hydroxyl groups. Such a molecule can be introduced into the bacterial cell whereupon these nicks are repaired.

Another major disadvantage of joining DNAs at restriction sites is often encountered when the interest lies in either cloning a large polypeptide coding sequence or a large segment of chromosomal DNA which contains several restriction sites. One early approach which was used to get around this problem was to clone partial *Eco*RI digests of chromosomal DNA in a bacterial plasmid [7]. This is quite a laborious technique because the partial digestion products and the

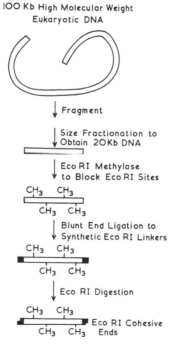

IOO Kb High Molecular Weight
Eukaryotic DNA

↓ Fragment

Size Fractionation to
Obtain 20Kb DNA

Eco RI Methylase
to Block Eco RI Sites

Blunt End Ligation to
Synthetic Eco RI Linkers

Eco RI Digestion

Eco RI Cohesive
Ends

Fig. 2.3 Addition of linkers to randomly fragmented chromosomal DNA. (Redrawn from [8]; copyright M.I.T.).

ligated *EcoRI* fragments have to be carefully sized in order to discriminate against clones containing oligomers of restriction fragments which were not originally adjacent to each other in the chromosome. There is no way around the application of such careful sizing steps, although some of the other shortcomings of this earlier method have been corrected in a general approach adopted by Maniatis and co-workers [8], who set out to build 'libraries' of cloned DNAs from genomes of higher organisms (Fig. 2.3). The distribution of restriction sites within the genome of an organism is not random, but is determined by the functional arrangements of nucleotide sequences, and so the cloning of DNA digested by an endonuclease such as *EcoRI* could lead to selective loss of DNA from pools of recombinants. In order to overcome this potential problem Maniatis *et al.* [8] generated randomly broken segments of chromosomal DNA by partial digestion with *Hae*III and *Alu*I. These enzymes recognise the tetranucleotide sequences GGCC and AGCT respectively, and cleave DNA to generate flush ends. Specific tetranucleotide sequences occur more frequently in DNA than specific hexanucleotide sequences, and so there is a high probability that any segment of DNA will contain cleavage sites for one of these two enzymes. This partially cleaved DNA is then fractionated by velocity sedimentation and 20Kb fragments are selected for cloning in a λ phage designed to accept *EcoRI* fragments. To achieve this, chemically synthesised oligonucleotides containing the *EcoRI*

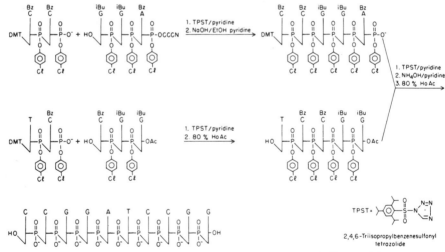

Fig. 2.4 Scheme for chemical synthesis of *Bam*HI linkers. (from [10]; copyright 1977 American Association for the Advancement of Science). The basis of the triester synthesis involves the phosphorylation of the 3′ hydroxyl group of a 5′-protected mononucleoside followed by condensation with the 5′ primary hydroxyl group of a 3′ protected nucleoside. The condensing agent is triisopropylbenzensulphonyltetrazolide (TPST) [13]. DMT is the 4.4 dimethoxytrytyl 5′ protecting group; CCCN is a β-cyanoethyl protecting group for the 3′ phosphate and acetate (Ac) is the protecting group for the 3′OH group. Details of the synthesis can be found in references [10, 11, 12, 13].

14

recognition sequence are added onto the flush ends generated by *Hae*III and *Alu*I. These 'linkers' are a self complementary sequence of ten residues which self-anneal to give a 'flush-ended' decamer which is joined onto the 20Kb DNA using T4 ligase. An *Eco*RI cohesive terminus can then be generated on the 20Kb DNA segments by *Eco*RI digestion. In order not to cleave at internal *Eco*RI sites within the chromosomal DNA at this step, the DNA is modified with the *Eco*RI methylase before the linkers are added.

Decanucleotide and dodecanucleotide linkers have been synthesised containing the recognition sequences of a number of restriction endonucleotides which generate cohesive ends [9, 10]. The scheme for the chemical synthesis of such a linker for the *Bam*HI site is shown in Fig. 2.4. Many of these linkers are now available commercially and they greatly increase the flexibility for cloning specific DNA fragments within the available vectors.

2.3 Joining DNA via homopolymeric tails

The experiments of Jackson *et al.* [14] using the techniques of Lobban and Kaiser [15] initiated the recombinant DNA furore. Their techniques were first successfully used by Wensink *et al.* [16] to produce autonomously replicating hybrid plasmids (Fig. 2.5). The method still relies upon restriction endonuclease cleavage to open the circular

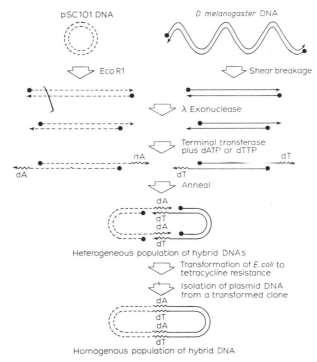

Fig. 2.5 The homopolymer tailing technique (from [16]; copyright M.I.T.).

plasmid DNA at a specific, non-essential site, but the donor DNA can be fragmented by a number of techniques and in this example *Drosophila melanogaster* DNA was fragmented by hydrodynamic shear to give fragments of 15–20Kb. The linear molecules to be joined are first treated with λ-exonuclease, which successively removes deoxymononucleotides from the 5′-phosphoryl termini of double stranded DNA, leaving single stranded 3′OH termini. Exposed 3′OH termini are good primers for calf thymus terminal transferase, which is used to add homopolymer blocks of deoxyadenylate and deoxythymidylate residues to the respective molecules. The two DNA preparations are then mixed and annealed, whereupon they will join via hydrogen bonding between their homopolymer tails. The original procedure of Lobban and Kaiser [15] then required the covalent sealing of such molecules by the concerted action of exonuclease III, DNA polymerase I and DNA ligase. This procedure was greatly simplified by Wensink *et al.* [16], who introduced the hydrogen bonded molecules directly into *E. coli* and selected transformants which were tetracycline resistant as determined by the plasmid vector pSC101. Roychoudhury *et al.* [17] have described a further simplification of this procedure whereby, using cobalt rather than magnesium as a divalent cation for the terminal transferase reaction, it is possible to add the homopolymeric tails to DNA molecules which have not been treated with λ exonuclease.

The advantages of this joining method for cloning chromosomal DNA are that one cannot get indiscriminate joining of DNA segments which originate from non-contiguous regions, and that it can be applied

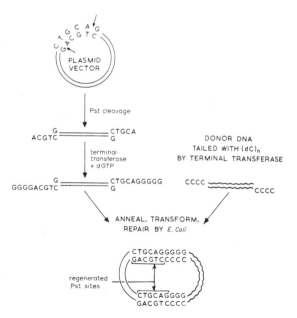

Fig. 2.6 Regeneration of the *Pst*I site by dG:dC tailing.

to randomly broken DNA segments. Furthermore, vector plasmid DNAs that have homopolymeric single strand extensions cannot self-cyclise, and consequently almost all of the bacteria transformed by DNA joined in this manner contain hybrid molecules. One disadvantage of the procedure is that it is difficult to cleave the cloned segment of DNA from the vector DNA in the hybrid plasmid. This is sometimes possible by Sl nuclease digestion, since Sl cleaves single stranded nucleic acids and will attack breathing duplexes in AT-rich regions.

Polydeoxyguanylate and polydeoxycytidylate tails can also be added to DNA molecules following these same procedures. If poly dG tails are added onto a plasmid vector cleaved with *Pst*I, the poly dC-tailed DNA cloned in such a plasmid can often be removed by *Pst*I cleavage, since this joining procedure regenerates the *Pst* I site (Fig. 2.6).

References

[1]   Arber, W. (1965), *Ann. Rev. Microbiol.* **19**, 365.
[2]   Meselson, M., Yuan, R., Heywood, J. (1972), *Ann. Rev. Biochem.* **41**, 447.
[3]   Roberts, R. J. (1980), *Nucleic Acids Res.* **8**, r63.
[4]   Mertz, J. E. and Davis, R. W. (1972), *Proc. natn Acad. Sci. USA* **69**, 3370.
[5]   Lehman, I. R. (1974), *Science* **186**, 790.
[6]   Sgaramella, V. and Khorana, H. G. (1972), *J. mol. Biol.* **72**, 493.
[7]   Glover, D. M., White, R. L., Finnegan, D. J. and Hogness, D. S. (1975), *Cell* **5**, 149.
[8]   Maniatis, T., Hardison, R. C., Lacy, E., Lauer, J., O'Connel, C., Quon, D., Sim, G. K. and Efstratiadis, A. (1978), *Cell* **15**, 687.
[9]   Bahl, C. P., Marians, K. J., Wu, R., Stavinsky, J. and Narang, S. (1977), *Gene* **1**, 81.
[10] Scheller, R. H., Dickerson, R. E., Boyer, H. W., Riggs, A. D. and Itakura, K. (1977), *Science* **196**, 177.
[11] Itakura, K., Natagiri, N., Bahl, C. P., Wightman, R. H. and Narang, S. A. (1975), *J. Am. chem. Soc.* **97**, 7327.
[12] Katagiri, N. Itakura, K. and Narang, S. A. (1975), *J. Am. chem. Soc.* **97**, 7332.
[13] Stavinski, J., Hozuni, T. and Narang, S. A. (1976), *Can. J. Chem.* **54**, 670.
[14] Jackson, D. A., Symons, R. M. and Berg, P. (1972), *Proc. natn Acad. Sci. USA.* **69**, 2904.
[15] Lobban, P. and Kaiser, A. D. (1973), *J. mol. Biol.* **78**, 453.
[16] Wensink, P. C., Finnegan, D. J., Donnelson, J. E. and Hogness, D. S. (1974), *Cell* **3**, 315.
[17] Roychoudhury, R., Jay, E. and Wu, R. (1976), *Nucleic Acids Res.* **3**, 101.

3 Plasmid vectors

A plasmid vector should have a marker function which enables the selection of bacteria carrying the plasmid, and should also have single restriction endonuclease cleavage sites in regions non-essential for DNA replication or for the marker function. Most vectors are non-conjugative plasmids and so do not bring about the transfer of DNA from one cell to another. This is a desirable property with respect to the containment of any potentially hazardous recombinant DNA molecule within a single microorganism. Conjugative plasmids such as the *E. coli* sex factor F, on the other hand, are capable not only of their own transmission from one cell to another, but also of transferring chromosomal markers at high frequency if they are integrated into the bacterial chromosome. They can also mobilise DNA to which they are not covalently joined and this includes co-existing non-conjugative plasmids. In this Chapter I will trace the modification of several non-conjugative plasmid vectors to give more effective cloning vehicles which can no longer be effectively mobilised during conjugation.

3.1 pSC101
This plasmid is derived from a self-transmissible plasmid R6-5, itself a derivative of the 97Kb R6 plasmid. R6 shares about 50Kb homology with F in that region which includes the *tra* operon, those genes necessary for the conjugal transfer of DNA [1]. R6 carries genes which determine resistance to streptomycin, sulphonamides, chloramphenicol, kanamycin and tetracycline. The R6–5 plasmid has lost the ability to confer tetracycline resistance but retains all other drug resistance markers. pSC101 arose from an experiment in which R6-5 DNA was hydrodynamically sheared and then used to transform *E. coli* to tetracycline resistance [2]. The 9Kb pSC101 plasmid was isolated from one such transformant. There is no satisfactory explanation for the genesis of pSC101. It is possible that it arose from the cyclisation of a segment of R6–5 DNA and the concomitant activation of the 'dormant' tetracycline resistance gene. Alternatively, pSC101 might have co-existed with the R-factor as a small sheer resistant supercoiled DNA molecule.

pSC101 was the first effective cloning vehicle to be used for cloning eukaryotic DNA [3]. It has the advantage of a single *Eco*RI site at which DNA can be inserted without affecting either the replicative properties or drug resistance marker of the plasmid. It has the disadvantage of being under stringent replicative control with only 1–2 copies of the plasmid per cell. Consequently the yields of plasmid

DNA from cells carrying pSC101 are low by comparison with the plasmid vectors that are in current use.

3.2 ColE1

This problem of low plasmid yield with pSC101 does not occur with the ColE1 plasmid, which is present in about 20 copies per cell. The copy number of this plasmid can be further increased to between 1000 and 3000 copies per cell by addition of chloramphenicol to a log-phase culture. Under these conditions the chromosomal DNA stops replicating, whereas the plasmid DNA continues to replicate and eventually makes up about 50% of the cellular DNA [4]. ColE1 is one of a series of plasmids which determine the production of the antibiotic proteins, the colicins. Each colicin has a characteristic mode of action: colicins E1 and K inhibit active transport [5, 6], colicin E3 inhibits protein synthesis by cleaving 16S rRNA [7], and colicin E2 promotes the degradation of DNA [8]. In all cases the end result is cell death, if that cell does not carry the colicinogenic plasmid. Cells carrying a Col plasmid are immune to the effects of the colicin specified by that plasmid. This forms the basis of a selection system for cells transformed by Col plasmids. The selection system has to be applied with care, however, since cells resistant to colicins arise spontaneously at quite a high frequency in a bacterial population. Nevertheless, the advantage of the high copy number of ColE1 plasmids outweighs this difficulty, and ColE1 has found extensive use as a cloning vehicle [9]. Foreign DNA may be inserted into the single *Eco*RI site without affecting either the replication properties or the plasmid's ability to confer colicin immunity, although the ability to produce the colicin protein is lost.

ColE1 is a non self-transmissible plasmid, but it can be mobilised by self-transmissible plasmids. 'Mini'-derivatives of ColE1 have been developed which are less readily mobilised and so are more readily acceptable as 'safe' vectors. There are two plasmids in the literature which have acquired the name 'mini-ColE1', and both replicate like ColE1 to give high copy numbers in the presence of chloramphenicol. One of these arose as a *trp*⁻ revertant from a recombinant plasmid of ColE1 carrying an *Eco*RI fragment with the *E. coli trp* genes [9]. It is a 3.4Kb plasmid (pVH51) which shows 3Kb homology with the 6.5Kb ColE1 genome [10]. The second plasmid was derived from the naturally occurring plasmid pMB1, which has extensive homology with ColE1 and which encodes the *Eco*RI restriction and modification enzymes and also determines resistance to ampicillin. The pMB1 plasmid was digested by *Eco*RI under conditions in which the enzyme only recognises the central four nucleotides of the *Eco*RI recognition site (*Eco*RI* conditions), and so cleaves at many more sites than does *Eco*RI. The resulting fragments still have the *Eco*RI AATT cohesive termini and so can be religated and reintroduced into *E. coli*. It is then possible to select recombinants which have lost *Eco*RI* fragments carrying markers from the original plasmid. One such colicin E1 immune ampicillin-sensitive clone contained the 2.7Kb plasmid pMB8 which is homologous to that same region of ColE1 present in pVH51 [11].

3.3 Derivatives of ColE1 which contain drug-resistance markers

In order to improve methods for selecting transformants, drug resistance genes have been introduced into ColE1 or its derivatives. Most drug resistance markers are carried on translocatable genetic elements (transposons). The drug resistance gene is characteristically flanked by homologous sequences [12]. These flanking sequences can be oriented in the same direction with respect to the drug resistance gene or in opposite orientation. In the latter case they form a characteristic snap-back stem and loop structure when visualised in the electron microscope following denaturation and brief annealing. Such drug resistance genes can undergo *rec*A independent translocation from one replicon to another. Deletions which extend from within the transposon into the externally flanking sequences are frequently generated upon transposition [13]. For such events to occur within a cloning vector is clearly disadvantageous, since they result in the loss of the selectable marker and potential loss of the cloned DNA sequence. The possible transfer of DNA from one replicon to another is furthermore undesirable in terms of the biological containment of potentially hazardous recombinant DNA molecules. Many of the early vectors carrying drug resistance markers did contain intact translocatable elements, but in the development of the vectors deletions have been generated within the repetitive elements which flank these genes and in such cases the elements are no longer capable of translocation.

One of the first derivatives of ColEl to carry a drug resistance marker, pML2, was an *in vitro* recombinant between *Eco*RI-cleaved ColE1 and an *Eco*RI fragment which carries the gene for kanamycin resistance. This *Eco*RI fragment originated in the R6–5 plasmid and was transferred first into pSC101 [14] and subsequently into ColE1 [9]. pML2 has therefore two cleavage sites for *Eco*RI and so it is difficult to use as a vector for *Eco*RI fragments. The problem was overcome by deleting sequences around one of these *Eco*RI sites, using a technique for *in vitro* mutagenesis which will be described in Chapter 7 [15]. One such single-site derivative, pCR1, has a deletion of 25–30 nucleotide pairs around one *Eco*RI site.

An alternative way of introducing drug resistance into ColE1 is by the *in vivo* translocation of such an element. The gene for ampicillin resistance, for example, is carried on the transposon Tn1 (formerly TnA). So *et al.* [16] were able to translocate TnA from an R plasmid (R1 *drd* 19) into ColE1 by cocultivation of two plasmids in the same bacterial cells. One such recombinant plasmid, pSF2124, contains genes for both colicin biosynthesis and ampicillin resistance. The plasmid has single sites for *Bam*H1 and *Eco*RI into which foreign DNA may be inserted without affecting the ampicillin resistance marker.

The next development in plasmid cloning vectors involved the incorporation of two drug resistance markers in such a way that one of these genes could be inactivated by the insertion of a foreign DNA segment. Bacteria containing recombinant plasmids can then be identified by virtue of their resistance to one drug, but sensitivity to the other. The hybrid plasmids made by Hamer and Thomas [17] between pSC101 and pSF2124 (pGM706) and between pSC101 and pML21 (pGM16) are examples of this plasmid type. These plasmids are, however, very large and contain many genes which are clearly redundant in the role of the cloning vector. They have not found general use since they were superceded by an elegant set of plasmids derived from the mini-ColE1 plasmid pMB8 (see Fig. 3.1).

The tetracycline resistance gene was first introduced into pMB8 by ligating EcoRI linears of this plasmid with fragments from an EcoRI* digest of pSC101. One recombinant plasmid arising from this experiment was the 5.3Kb pMB9 which combines the advantageous replicative properties of ColE1 with the Tc^r marker. This plasmid has a

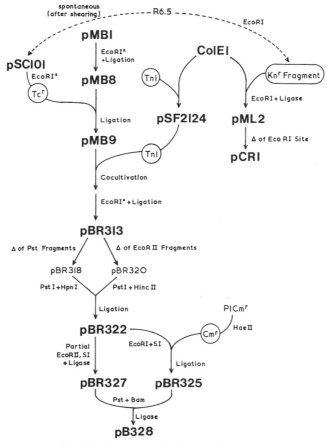

Fig. 3.1 Genealogy of some plasmid vectors.

single *Eco*RI site and has been extensively used as a vector. pMB9 has also single sites for *Hin*dIII, *Bam*H1 and *Sal*I, but insertion of foreign DNA into these sites inactivates the Tc^r marker. In order to be able to use these sites for cloning and still be able to apply strong selection for drug resistance, the gene for ampicillin resistance (Ap^r) has been introduced into pMB9. This was achieved by cultivating pMB9 with pSF2124 so that Tn1 could undergo transposition into pMB9. A number of plasmids which determined resistance to both antibiotics were isolated from such experiments. In order to remove the additional cleavage site for *Bam*H1 introduced on Tn1, one of these *in vivo* recombinant plasmids was subjected to *Eco*R1* digestion and religation of the resulting fragments. This DNA was again introduced into *E. coli*, and cells were selected which were

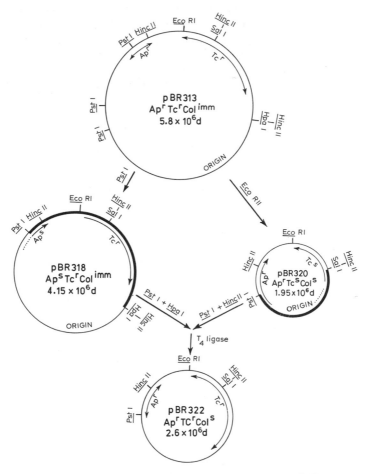

Fig. 3.2 Construction of pBR322 from pBR313 (from [19]).

resistant to both tetracycline and ampicillin. One of the transformants harboured the plasmid pBR313, which has only a single *BamH*1 site in the *Tc^r* gene. Foreign DNA can be inserted into this site or the *Hind*III and *Sal*I sites, thereby inactivating the *Tc^r* gene but leaving the *Ap^r* gene functional (see Fig. 3.2). The deletion of sequences containing the *BamH*1 site of Tn1 has the additional advantage of preventing translocation of the *Ap^r* to other episomes [18].

The gene for ampicillin resistance contains a site for *Pst*I, and so cloning into this site results in recombinant plasmids which would give an *Ap^sTc^r* phenotype. This is not practical in pBR313 since it contains two other *Pst* sites. These sequences have been deleted in the plasmid pBR322. The construction of pBR322 involved firstly deletion of two *Pst*I fragments from pBR313 to form the *Ap^sTc^r* plasmid pBR318. In a second experiment the *Eco* RII fragments were deleted to form the *Ap^rTc^s* plasmid pBR320 . In the final step segments of these two plasmids were recombined *in vitro*. [19] (Fig. 3.2). pBR322 has been completely sequenced and is currently the most extensively used plasmid vector.

The gene for chloramphenicol resistance (*Cm^r*) has also been introduced into pBR322 on a *Hae*II fragment from a P1 phage. This fragment was ligated to *Eco*RI-generated linears of pBR322 which had been treated with S1 nuclease to remove the *Eco*RI cohesive ends. This 'blunt end ligation' step (see Chapter 2) effectively destroys these two

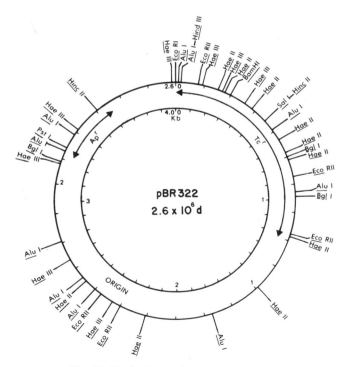

Fig. 3.3 Physical map of pBR322. (from [20]).

*Eco*RI recognition sites and so the only *Eco*RI site remaining in this plasmid, pBR325, is located in the *Cm^r* gene. Inactivation of the *Cm^r* gene can therefore be used as a means of identifying *in vitro* recombinant plasmids in which the foreign DNA has been inserted into the *Eco*RI site [21].

3.4 The biological containment of a plasmid vector system

In order to contain cells carrying potentially hazardous cloned genes, steps have been taken firstly to prevent transfer of the plasmid from one strain to another, and secondly to introduce mutations into the host bacterium so that it can only grow in laboratory conditions. Like the other 'mini' derivatives of ColE1, the pBR322 plasmid can not be mobilised by a transmissible plasmid [19]. pBR322 can, however, be mobilised if a third plasmid, for example ColK, is present [22]. It is proposed that in this case the ColK plasmid provides a trans-acting 'mobility protein' which interacts with a specific DNA sequence (the '*bom*' site) on pBR322. The gene for the 'mobility protein' has been deleted from pBR322 but not its site of action. Derivatives of pBR322 have been constructed in which this site is deleted. In one such plasmid, pBR327, a 1089 residue *Eco*RII fragment has been deleted. This was achieved by partial *Eco*RII digestion, selection of the appropriate partial digestion product by gel electrophoresis, removal of protruding single stranded termini with S1 nuclease and religation of the blunt ends using T4 ligase. This also removes recognition sites for *Pvu*II and *Bal*I from the molecule. (A similar '*bom*' derivative of pBR322, pAT153, has been constructed by Twigg and Sherratt [23]). The *Pst*I/*Bam*HI fragment from pBR327 which contains this 1089 residue deletion has been ligated to the *Pst*I/*Bam*HI fragment of pBR325 which carries the *Cm^r* gene to produce a recombinant pBR328. The plasmid pBR328, like pBR325, carries *Cm^rAp^rTc^r* genes but now has single sites for *Pvu*II and *Bal*I in addition to *Eco*RI in the *Cm^r* gene.

A bacterial strain which has found general use in cases where biological containment is necessary is $\chi 1776$ [24]. This strain has mutations which make its growth absolutely dependent upon the presence of diaminopimelic acid and thymidine in the culture medium. Diaminopimelic acid (DAP) is not found within mammalian intestinal tracts, and so the strain will undergo DAP-less death if ingested. The strain also shows increased sensitivity to ultraviolet irradiation as a result of removal of one of the genes responsible for DNA repair, *uvrB*, and it is very sensitive to bile salts and ionic detergents. The strain is consequently difficult to handle in the laboratory and scrupulously clean glassware has to be used. It is possible, however, to transform the strain with efficiencies comparable to those obtained with other commonly used recipient strains.

3.5 Selection of plasmids containing specific nucleotide sequences

DNA molecules can be introduced into cells rendered competent by treatment with Ca^{++} ions [25]. In practice, *in vitro* recombinant

DNA molecules are introduced into cells in this way and the cells which acquire the plasmid are selected by their expression of the plasmid marker function. In an experiment in which a set of DNA segments from a complex genome have been recombined with a plasmid vector there will result a heterogeneous population of bacterial colonies, most of which will contain a unique cloned segment of the complex genome. The power of the basic cloning technique was realised in the development of a method which allows the selection of colonies which contain specific sequences [26]. The heterogeneous population of bacterial colonies can be replica plated onto an agar plate overlaid with a nitrocellulose filter. Nutrients diffuse through the filter and permit the growth of the replicated colonies. The colonies are lysed *in situ* on the filter in the position of the colony. The filter is then incubated with a radiolabelled probe, for example rRNA or a cDNA transcript of an mRNA, which will hybridise to complementary recombinant plasmid DNA present in some proportion of the transformants. Those colonies which hybridise can subsequently be localised by autoradiography and then picked from the master plate. In this way it has been possible to select many specific eukaryotic genes cloned in *E. coli* (See Chapter 6 for examples).

References

[1] Sharp, P. A., Sohen, C. N. and Davidson, N. (1973), *J. mol. Biol.* **75**, 235.
[2] Cohen, S. N. and Chang, A. C. Y. (1973), *Proc. natn Acad. Sci.* **70**, 1293.
[3] Morrow, J. F., Cohen, S. N., Chang, A. C. Y., Boyer, H. W., Goodman, H. M. and Helling, R. B. (1974), *Proc. natn. Acad. Sci. U.S.A.* **71**, 1743.
[4] Clewell, D. B. and Helinski, D. R. (1972), *J. Bact.* **110**, 1135.
[5] Fields, K. L. and Luria, S. E. (1969), *J. Bact.* **97**, 57.
[6] Fields, K. L. and Luria, S. E. (1969), *J. Bact.* **97**, 64.
[7] Boon, T. (1972), *Proc. natn Acad. Sci. U.S.A.* **69**, 549.
[8] Ringrose, P. (1970), *Biochim. biophys. Acta.* **213**, 320.
[9] Hershfield, V., Boyer, H. W., Yanofsky, C., Lovett, M. A. and Helinski, D. R. (1974), *Proc. natn Acad. Sci. U.S.A.* **71**, 3455.
[10] Hershfield, V., Boyer, H. W., Chow, L. and Helinski, D. R. (1976), *J. Bact.* **126**, 447.
[11] Rodriguez, R. L., Bolivar, F., Goodman, H. M., Boyer, H. W. and Betlach, M. In *ICN-UCLA Symposium* **5**, Academic Press.
[12] Kleckner, N. (1977), *Cell* **11**, 11.
[13] Ross, D. G., Swan, J. and Kleckner, N. (1979), *Cell* **16**, 721.
[14] Cohen, S. N., Chang, A. C. Y., Boyer, H. W. and Helling, R. B., (1973), *Proc. natn. Acad. Sci. USA.* **70**, 3240.
[15] Covey, E., Richardson, D. and Carbon, J. (1976), *Mol. gen. Genet.* **145**, 155.
[16] So, M., Gill, R. and Falkow, S. (1975), *Mol. gen. Genet.* **142**, 239.
[17] Hamer, D. H. and Thomas, C. A. (1976), *Proc. natn. Acad. Sci. USA* **73**, 1537.
[18] Bolivar, F., Rodriguez, R. L., Betlach, M. C. and Boyer, H. W. (1977), *Gene* **2**, 75.
[19] Bolivar, F., Rodriguez, R. L., Greene, P. J., Betlach, M. C., Heyneker, H. L., Boyer, H. W., Crosa, J. H. and Falkow, S.: (1977), *Gene*, **2**, 95, and:

[20] (1977), in: *DNA Insertion Elements, Plasmids, and Episomes* (Bukhari, A. I., Shapiro, J. A., Adhya, S. L., eds.), pp 686–7, Cold Spring Harbour Laboratory, U.S.A.

[21] Bolivar, F. (1978), *Gene*,4, 121.

[22] Young, I. G. and Poulis, M. I. (1978), *Gene*, 4, 175.

[23] Twigg, A. J. and Sherratt, D. (1980) *Nature*, 283, 216

[24] Curtiss, III, R., Pereira, D. A., Clark, J. E., Hsu, J. C., Goldsmidt, R., Hull, S. I., Moody, R., Maturin, L. and Inoe, M. (1976), In: *Proceedings of the Tenth Miles Symposium*—Raven Press, New York.

[25] Mandell, M. and Higa, A. (1970), *J. mol. Biol.* 53, 159.

[26] Grunstein, M. and Hogness, D. S. (1975), *Proc. natn Acad. Sci. USA* 72, 3961.

4   Bacteriophage λ vectors

4.1 The biology of phage λ

Bacteriophage λ has about 50 genes in its 49Kb genome, only about half of which are essential for its lytic growth. Before describing λ vectors in which this non-essential DNA can be replaced by foreign DNA, I will give a simplified description of the life cycle of the bacteriophage. More comprehensive reviews can be found elsewhere [1, 2].

Adsorption of phage λ involves an interaction between the tip of the phage tail with a component of the outer cell membrane of *E. coli*. Following penetration, the linear double stranded DNA molecule cyclises at the *cos* sites, which are single stranded cohesive ends consisting of mutually complementary sequences of 12 residues [3]. Early in the infection cycle these circular DNA molecules replicate as theta (θ) forms. The replication is bidirectional [4]; it originates between the genes *CII* and *O* and requires the activity of the phage genes *O* and *P*. Later in infection there is a switch towards a rolling circle mechanism which produces long concatameric molecules composed of several linearly arranged genomes [5]. The switch is brought about by the action of the phage *gam* gene product, which probably acts by inhibiting the *E. coli recBC* nuclease [6]. The maturation of these concatomers to unit length molecules occurs during packaging by the action of a nuclease at the *cos* sites [7]. This process, which also requires the presence of four head proteins, will be discussed later in this Chapter.

The bacteriophage genome has an alternative mode of propagation whereby it becomes stably integrated into the host chromosome and is replicated along with the bacterial chromosome with all but one of its genes repressed. Integration occurs as a result of recombination between the phage attachment site (*att*) and a partly homologous site *attλ* on the *E. coli* chromosome. Integration, which requires expression of the phage gene *int*, is a reversible process. Prophage excision requires activity of a phage gene *xis* in concert with *int*.

The normal $att\lambda$ site maps between the genes for galactose utilisation (*gal*) and biotin biosynthesis (*bio*). Abnormal excisions of the prophage can result in the incorporation of genes from one or other of these operons into the phage genome, with the concomitant deletion of some phage DNA. Depending upon the extent of this deletion, such transducing phage may or may not be defective for vegetative growth or lysogenisation. If the normal $att\lambda$ site is deleted from the *E. coli* chromosome then the phage can integrate at secondary attachment sites. Abnormal excisions of prophage from these other regions of the genome can result in the transducing phage carrying a number of other *E. coli* genes.

The decision between the lytic and lysogenic events is dependent upon the interactions of two proteins, the *CI* gene product and the *cro* gene product, with the two λ operators. The following is a highly simplified account of the sequence of gene expression for the two responses, and serves to introduce the reader to various promotors in the genome, some of which can be used to enhance the expression of cloned DNA sequences. If the two operators are free of repression then leftward and rightward transcription can take place on opposite strands from the two promotors P_L and P_R. Leftward transcription terminates at t_L to give a 12S mRNA which encodes the N gene product. The majority of the rightward transcripts terminate at t_{R1} to give the 7S mRNA for the *cro* gene product [8]. There is some readthrough to t_{R2} to give transcripts of the *CII, O* and *P* genes. Readthrough at each of these three terminators is enhanced by the N gene product [9, 10], and this leads to synthesis of the *CIII* gene product and those gene products involved in recombination from leftward transcription. It also results in elevated levels of the *CII, O, P* gene products and also of the *Q* gene product from rightward transcription. At this stage the *cro* gene product has reached such a concentration that it will shut down transcription from P_L and P_R by binding to the operators. The *Q* gene product meanwhile activates very efficient transcription from P_R' to ensure large quantities of head and tail proteins (Fig. 4.1).

The products of the *CII* and *CIII* can stimulate transcription from P_{re}, the promoter for the establishment of lysogeny resulting in the synthesis of the *CI* gene product. This binds tightly to P_L and P_R, and if sufficient repressor is made before late transcription can get established then all the genes of the phage will be repressed and the genome will be integrated into the chromosome by the *int* gene product. The repression of the *CII* and *CIII* genes in its turn leads to the cessation of transcription from P_{re}. The leftward and rightward operators each contain three binding sites for *CI* repressor, and all of these are filled when the repressor concentration is high. Upon division of the lysogenised bacterium, the sites empty in order of their binding affinity. When only one of these sites is bound to repressor, then the P_{rm} promoter (repression maintenance) is activated and more *CI* repressor can be synthesised [11] (Fig. 4.1). By incorporating a temperature-sensitive mutation in the *CI* gene, such that the polypeptide is inactivated at

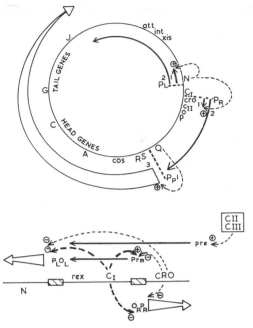

Fig. 4.1 The major transcription circuits of phage λ.

high temperature (usually 42° to 45°), it is possible to construct lysogens which can be conveniently induced by shifting the temperature of the culture. In nature, induction usually occurs in response to a mutagen such as UV light, which is thought to act by inducing the host *rec*A protein which can proteolytically cleave and so inactivate the *CI* repressor [12]. This allows the initiation of leftward and rightward transcription from P_L and P_R. Phage λ is one member of the family of lambdoid phages, all of which are active in *E. coli* but which display different immunities [13]. Immunity to superinfection is a characteristic of lysogenic cells resulting from the presence of *CI* repressor in their cytoplasm. The λCI gene product will not, however, repress phage 434 or phage 21, for example, since the operators of these latter phages will not bind the λ repressor. These phage genomes do, however, have considerable regions of sequence homology, and so it is possible to construct recombinant λ phage which carry the immunity region (the operators, *CI* and *cro* genes) of phage 434 for example. Such a phage is called λimm[434] and we shall see that such hybrid phage provide a useful means of varying the number of restriction sites for a given enzyme in λ phage vectors.

4.2 Phage vectors
Before phage λ could be used as a cloning vehicle, it was first necessary to eliminate from its genome some of the restriction sites for the enzymes commonly used for cloning. Derivatives of λ were initially constructed

with a reduced number of EcoRI sites, and with the remaining EcoRI sites in a non-essential region of the genome such that phage DNA could be completely cleaved with EcoRI, and foreign DNA inserted into this region. The strategy was first to remove all the sites and then replace the desired ones by *in vivo* genetic recombination. Rambach and Tiollais [14] and Murrary and Murrary [15] set out with phage which had deletions resulting in the loss of the two left most sites (sites 1 and 2). Thomas *et al.* [16] used a transducing phage which had also lost site 3 as a result of a *bio* substitution, and which contained a sequence duplication in its left arm, since DNA will only be packaged if its length is between 79% and 109% of wild type λ DNA. Mutants lacking the remaining sites were selected by cycling the phage between hosts containing, and hosts lacking, the EcoRI restriction-modification system. This was continued for several cycles until the efficiency of plating indicated that the DNA could no longer be restricted. The resulting phage which completely lacked EcoRI sites were crossed with phage containing all the EcoRI sites, and recombinants selected which contained only sites 1 and 2, or 1, 2 and 3. These phage can be cleaved with EcoRI and fragments of foreign DNA inserted between the left and right phage arms. Phage λ vectors have also been constructed for the restriction fragments produced by HindIII [17]. The wild type phage DNA has six sites for this enzyme. Deletion of the fragment between EcoRI sites 1 and 2 removes 2 HindIII sites. Alternatively, phage containing the b538 deletion can be used, since these lack HindIII sites 1, 2 and 3. The substitution of imm^{21} for $imm\lambda$ removes HindIII sites 4 and 5. Site 6 is lost by the substitution of ϕ 80 DNA in this region of the genome and this has been achieved by selecting for *in vivo* recombinants around the Q gene (Fig. 4.2).

Hybrid phage have been constructed which contain the CI gene of phage 434; this immunity region contains single cleavage sites for HindIII and EcoRI. A phage from this series of λimm^{434} hybrid phage vectors is illustrated in Fig. 4.2 [17]. The insertion of foreign DNA inactivates the CI gene, resulting in clear plaque morphology. This enables *in vitro* recombinant phage to be distinguished from parental vector phage which can lysogenise and so give turbid plaques (Fig. 4.3). Similar insertion vectors are to be found in the 'Charon' series developed by Blattner *et al.* [18] (Fig. 4.2). Charon 6 and Charon 7 each contain the imm^{434} region, although whereas Charon 7 will serve as an insertion vector for both HindIII and EcoRI, Charon 6 will only serve as an effective insertion vector for EcoRI fragments, since it retains a second HindIII site (site 6). There are alternative insertion vectors in the Charon series: insertion into the single EcoRI site of the *red* gene of Charon 12 should make the phage unable to grow on *E. coli* with mutant DNA polymerase I. Similarly, insertion of DNA into the single EcoRI site of Charons 2 or 16 should inactivate the lac Z gene carried on these phage, and this event should be detectable on colour indicator plates (see also below).

The formation of plaques by *in vitro* recombinant phage can be made to depend upon the reconstitution of a molecule of a certain size, since the DNA will only be packaged into mature phage if its length is within the packaging limits. The amount of DNA that can be cloned in phage λ can therefore be increased by using a phage that has two sites for a restriction endonuclease bordering a non essential region. The non-essential phage DNA serves to assist propagation of the vector and is subsequently replaced with foreign DNA. The packaging limitations ensure that the viable products of *in vitro* recombination must either have had this central fragment reincorporated into the genome or replaced by foreign DNA. Deletion of this fragment would generate a DNA molecule too small to be packaged.

In the replacement vector, λgt. λC, of Thomas *et al.* [16], the replacement fragment is that between *Eco*RI sites 2 and 3 (fragment C). It contains *att*, *int* and *xis* and gives the phage the capability of forming stable lysogens. When fragment C is replaced by foreign DNA, however, the phage becomes integration defective. Murray *et al.* [17] have described a series of replacement vectors for *Eco*RI or *Hind*III fragments which simplify the recognition of recombinants. One such phage, λNM781, has a replaceable *Eco*RI fragment which carries the gene, *sup*E, for a mutant tRNA of *E. coli* (Fig. 4.2). This gene is active when the *Eco*RI fragment is oriented in either direction in the λ chro-

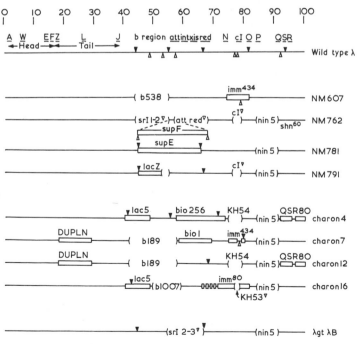

Fig. 4.2 Maps of some λ vectors. Dark triangles above the map represent *Eco*RI sites and open triangles below the map *Hind* III sites.

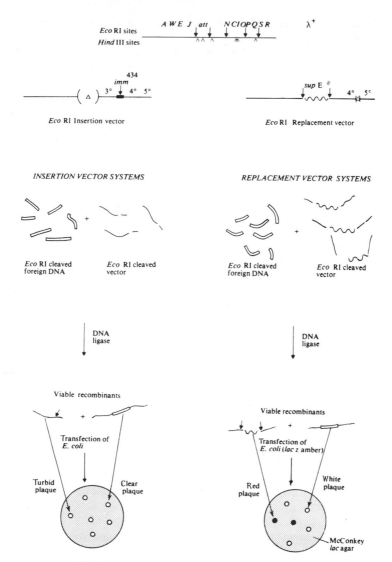

Fig. 4.3 Insertion vectors and replacement vectors.

mosome. The phage is recognised by the suppression of an amber mutation in the *lacZ* gene of the bacterial host, either as red plaques on lactose/MacConkey agar or as blue plaques on agar containing 4-bromo-5-chloroindol-3-yl β-D-galactoside (XG). The *in vitro* recombinant phage give colourless plaques on both these indicators (see Fig. 4.3). A similar replacement vector for *Hind*III fragments is described, λNM762, containing a *Hind*III fragment for the *supF* gene.

An alternative vector has most of the *lacZ* gene as a replaceable

*Eco*RI fragment (λNM791) which can be recognised by allelic complementation of a suitable *lac⁻* indicator strain. Many Charon phages also carry the *lac5* substitution which contains the βgalactosidase gene together with its operator and promoter. In the replacement vector Charon 4, for example, most of the *lac* DNA is on a single *Eco*RI fragment, and this, together with the adjacent *Eco*RI fragment which contains a *bio* substitution, can be replaced by foreign DNA. The parental phage will give dark blue plaques when plated on medium containing the βgalactosidase substrate, XG. When the *Eco*RI fragment containing *lac5* is replaced with foreign DNA, colourless plaques are produced. If the fragment is rearranged in Charon 4 as a result of *in vitro* recombination, or if in the case of Charon 16 a foreign sequence is inserted into the *lac* gene, then the plaques will be colourless on *lac⁻* indicator bacteria but pale blue on *lac⁺* strains. In the latter case, the increased gene dosage of *lac* operator produced by phage growth is thought to titrate out the cell's *lac* repressor, so causing some derepression of the bacterial *lac* operon [18].

4.3 The late genes—Their exploitation in cloning vectors

4.3.1. *Phage assembly*
The genes for the late proteins are effectively transcribed from P'_R, providing it has been activated by the Q gene product. The late genes include S and R, which encode proteins required for host lysis and some 20 genes for head and tail proteins. *In vitro* complementation studies [19, 20] have played an important part in elucidating the process of phage maturation. Head and tail assembly occur separately, so that phage with mutations in head genes can only make tails, and phage with mutations in tail genes can only make heads. A mixture of extracts from these two classes of mutant enables viable phage particles to be assembled *in vitro*. This assay has been extended to encompass complementation in pairs of head mutants and has proved invaluable in establishing the sequence in which the phage maturation functions act. The products of genes E and D (pE and pD) are the major capsid proteins which account for 72% and 20% of the total head protein respectively. The other minor head components are: pB*, derived by proteolytic cleavage of the gene B product; X1 and X2, two similar polypeptides derived from the fusion of part of pE with pC; pF11 and pW, which are needed for the head to combine with a tail. A model for the sequential process of head assembling is shown in Figure 4.4. Small empty headed particles known as *petit lambda* (pλ), are produced in normal infections although differing forms of pλ are produced in infections with phage mutated in late genes. Prohead I structures accumulate with either $C⁻$ or $B⁻C⁻$ phage and in phage grown in *E. coli* with the *groE* mutation. These appear to contain a protein core which is lost in the transition to the prohead II structure. The cleavage of pB to pB* and the cleavage and fusion of pC and pE to give X1 and X2 requires the putative core protein pNu3. The products of genes A and D

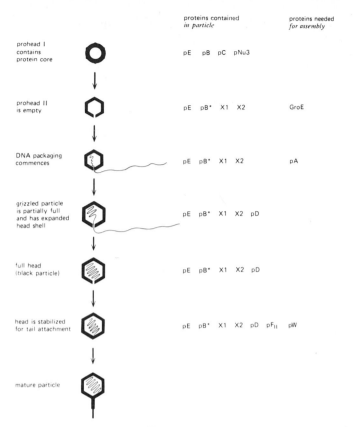

		proteins contained *in particle*	proteins needed *for assembly*
prohead I contains protein core		pE pB pC pNu3	
prohead II is empty		pE pB* X1 X2	GroE
DNA packaging commences		pE pB* X1 X2	pA
grizzled particle is partially full and has expanded head shell		pE pB* X1 X2 pD	
full head (black particle)		pE pB* X1 X2 pD	
head is stabilized for tail attachment		pE pB* X1 X2 pD pF$_{II}$ pW	
mature particle			

Fig. 4.4 Head assembly (from [2] Reprinted by permission of John Wiley & Sons, Ltd).

then act to initiate filling of the shell with DNA and the expansion of the shell to its mature size. The concatameric DNA molecules produced late in infection are cleaved to unit length by the action of pA at the *cos* site. Finally pW and pF11 are added in sequence to generate a structure that is able to bind mature tails [19, 20].

4.3.2. *Biological containment*

In order to enhance biological containment amber mutations have been introduced into a number of vectors. Three of the Charon phages have, for example, been certified for use in an EK2 vector-host system; that is to say they permit the lowering of physical containment requirements by one category. These phage are Charons 3A, 4A and 16A which differ from Charon 3, 4 and 16, from which they are derived, by the introduction of amber mutations in genes *A* and *B*. Infections resulting from these phage will therefore produce mature phage particles only in a suppressor strain of *E. coli* and not in wild type strains. The bacterial host strain, DP50supF, selected for use with the Charon phages, is a derivative of a Strain χ 1953 into which the *supF*58 (*su*III) mutation has been introduced. Like the strain χ 1776, described in the previous chapter, the growth of DP50supF is absolutely dependent upon the presence of diaminopimelic acid and thymidine in the culture medium. Charons 3A, 4A and 16A are furthermore confined to the lytic mode of interaction with the host cell by deletions in the immunity regions, and by the *nin* 5 deletion. Furthermore, Charon 4A and 16A recombinant molecules cannot lysogenise since they lack *int* and *att*.

Derivatives of the phage of Thomas *et al.* [16] have also been commonly used as EK2 vectors [21]. These derivatives have amber mutations in genes *W, E* and *S*. The *supF* suppressor is specifically required for the suppression of the mutant *S* gene in these phage, and so once again the DP50supF strain is a suitable host. The phage λgtWES. λB has a deletion of the *Eco*RI fragment between sites 2 and 3, and so it is *att*⁻ and *int*⁻, and cannot lysogenise. The replaceable *Eco*RI fragment in this case is phenotypically inert, and so recombinants with this phage cannot be recognised by their phenotype. This *Eco*RI fragment, however, contains the only cleavage sites for *Sst*I within the phage genome, and so the background of parental phage to recombinant phage can be greatly reduced by cleaving the vector DNA with *Eco*RI and *Sst*I before carrying out the ligation reaction.

4.3.3 *Increased recovery of recombinants by in vitro packaging*
The transfection of phage DNA into *E. coli* cells rendered competent by the procedure of Mandel and Higa [22] is an inefficient process giving between 10^5 and 10^6 plaques per μg of phage DNA (5×10^{-6} to 5×10^{-5} plaques per DNA molecule). Hohn and Murray [23] and Sternberg *et al.* [24] have developed an *in vitro* packaging system which can yield as many as 10^8 plaques per μg of DNA. In this system, the DNA to be packaged is mixed together with concentrated lysates from two cultures; one of cells which are undergoing infection with a phage whose maturation is blocked by an amber mutation in gene *D*, for example, and the other in which the infecting phage has a mutation, for example, in gene *E*. The two lysates complement each other *in vitro*, and encapsidate concatomers of λ DNA to form mature phage. In practice the packaging extracts are made from *E. coli* strains lysogenised by a phage with a temperature-sensitive repressor The phage infection is then initiated upon thermoinduction of the lysogen.

The *in vitro* packaging systems are a key requirement in the efficient construction and screening of 'libraries' of cloned segments from complex genomes. Most of the replacement vectors which I have described facilitate the recognition of the reconstituted parental phage. In order, however, to ensure a high proportion of recombinants it is desirable to purify the left and right arms of the vector away from the replaceable segment. This may be done by allowing the vector DNA to anneal through its cohesive ends, and then treating with DNA ligase to form covalently closed circles. Restriction-endonuclease cleavage will then generate a fragment consisting of the joined left

34

and right λ ends, which is easily separated from the replaceable segments by sucrose gradient sedimentation. This approach was used by Maniatis *et al.*, [25] in their construction of libraries of eukaryotic chromosomal segments (Chapter 2). Recombinant λDNA is then introduced into *E. coli* by the *in vitro* packaging process, and the resultant plaques can be screened for complementarity towards a given radiolabelled sequence by the nucleic acid hybridisation procedure of Benton and Davis [26]. The principle of the procedure is the same as that of colony hybridisation (See Chapter 2). A nitrocellulose filter is laid onto the *E. coli* lawn and phage are absorbed onto the filter from the plaques. DNA is released from the phage on the filter and denatured by alkali treatment. The filter is brought to neutral pH, so allowing single stranded DNA to stick to the filter in the position of the plaque. The filter can then be incubated with a radiolabelled probe for a specific nucleotide sequence from the gene of interest. The position of hybridisation of the probe is located by autoradiography and so the corresponding plaque can be picked from the master plate. This procedure permits the rapid screening of several hundred thousand plaques and so facilitates the isolation of single copy genes from complex eukaryotic genomes (see also Chapter 6).

The λ *cos* site has also been incorporated into plasmid DNA molecules [27]. These molecules can then be packaged *in vitro* and so introduced into *E. coli* with high efficiency. Such cloning vectors have been termed 'cosmids'. Since the same packaging limitations apply in the use of cosmids as with λDNA, and since in general cosmids are much smaller than λ vectors, they provide a cloning technique which favours the propagation of large segments of foreign DNA. In order to clone such segments in cosmids, the *in vitro* recombination reactions are best carried out at high DNA concentrations to favour the production of oligomers, which resemble the molecules produced late in phage infections. These oligomeric structures are cleaved at the *cos* site upon packaging. Once the molecule is injected from the phage head into the cell, it cyclises through the *cos* site and replicates as a plasmid.

References

[1] Hershey, A. D., ed (1971), 'The Bacteriophage Lambda' Cold Spring Harbor Laboratory, New York (1971).

[2] 'Gene Expression III' (1977), B. Lewin. John Wiley.

[3] Wu, R. and Taylor, E. (1971), *J. mol. Biol* **57**, 491.

[4] Inman, R. B. (1966), *J. mol. Biol.* **18**, 464.

[5] Bastia, D., Sneoka, N. and Cos, E. C. (1975), *J. mol. Biol.* **98**, 305.

[6] Enquist, L. W. and Skalka, A. (1973), *J. mol. Biol.* **75**, 185.

[7] Wang, J. C. and Kaiser, A. D. (1973), *Nat. N. B.* **241**, 16.

[8] Roberts, J. (1969), *Nature* **224**, 1168.

[9] Adhya, S., Gottesman, M. and De Crombrugghe, B. (1974), *Proc. natn Acad. Sci. USA.* **71**, 2534.

[10] Franklin, N. (1974), *J. mol. Biol.* **89**, 33.

[11] Ptashne, M., Backman, K., Humaynn, M. Z., Jeffrey, A., Maurer, R., Meyer, B. and Sauer, R. T. (1976), *Science* **194**, 156.
[12] Roberts, J. W., Roberts, C. W. and Mount, D. W. (1977), *Proc. natn Acad. Sci.* **74**, 2283.
[13] Simon, M. N., Davis, R. W. and Davidson, N. (1971), In 'The Bacteriophage Lambda' (Hershey, A. D., ed.), Cold Spring Harbor Laboratory, New York.
[14] Rambach, A. and Tiollais, P. (1974), *Proc. natn Acad. Sci. USA.* **71**, 3927.
[15] Murray, N. E. and Murray, K. (1975), *Nature* **251**, 476.
[16] Thomas, M., Cameron, J. R. and Davis, R. W. (1974), *Proc. natn Acad. Sci. USA.* **71**, 4579.
[17] Murray, N. E., Brammar, W. J. and Murray, K. (1977), *Mol. gen. Genet.* **150**, 53.
[18] Blattner, F. R., Williams, B. G., Blechl, A. E., Denniston-Thompson, K., Faber, M. E., Furlong, L. A., Grunwald, D. J., Kiefer, D. O., Moore, D. D., Schumm, J. W., Sheldon, E. L. and Smithies, O. (1977), *Science* **196**, 161.
[19] Hohn, T., Wurtz, M. and Hohn, B. (1976), *Phil. Trans. R. Soc.* **276**, 51.
[20] Casjens, S. and King, J. (1975), *Ann. Rev. Biochem.* p. 555.
[21] Leder, P., Tiemeier, D. and Enquist, L. (1977), *Science* **196**, 175.
[22] Mandel, M. and Higa, A. (1970), *J. mol. Biol.* **53**, 159.
[23] Hohn, B. and Murray, K. (1977), *Proc. natn. Acad. Sci. USA.* **74**, 3259.
[24] Sternberg, N., Teimeier, D. and Enquist, L. (1977), *Gene* **1**, 255.
[25] Maniatis, T., Hardison, R. C., Lacy, E., Lauer, J., O'Connel, C., Quon, D., Sim, G. K. and Efstratiadis, A. (1968), *Cell* **15**, 687.
[26] Benton, W. and Davis, R. W. (1977), *Science* **196**, 180.
[27] Collins, J. and Bruning, H. J. (1978), *Gene* **4**, 85.

5 Expression of cloned DNAs in *E. coli*

In order to achieve expression of a foreign gene in a bacterial cell it is necessary to put that gene under the control of the transcriptional and translational machinery of the host cell. The bacterium *E. coli* is a natural choice for such studies because of our detailed knowledge of the molecular biology of its gene expression. This chapter will examine the suitability of the most commonly used cloning vehicles as systems for expressing foreign genes.

5.1 Expression of DNA cloned in plasmid vectors

5.1.1 *Complementation of E. coli auxotrophs*
The power of this general approach was impressively demonstrated by Clarke and Carbon [1], who made 'banks' of cloned random

36

segments of *E. coli* DNA inserted into Co1E1 by the dA:dT tailing technique. They successfully isolated portions of the *ara* and *trp* operons by selecting from these 'banks' plasmids which would complement auxotrophic strains. As a variant on this direct method of selection which requires an auxotrophic strain that can be made competent for transformation, they used one pool of recombinant plasmids to transform a strain of *E. coli* harbouring an F plasmid [2]. The recombinant plasmids in their transformants could then be introduced into a number of F⁻ auxotrophic strains by F-mediated transfer in a replica-mating experiment. In this way many genes were isolated from this one bank. The potential of this approach for the selection of prokaryotic genes is clear. The use of F plasmids to promote the conjugal transfer of recombinant plasmids containing DNA from higher organisms was not, however, approved in the N.I.H. guidelines. It was feared that in the event of accidental ingestion of *E. coli* containing such plasmids there could be rapid colonisation of a variety of intestinal microorganisms by recombinant plasmids, some of which might carry a hazardous trait.

To date, however, *E. coli* auxotrophs have only been successfully complemented by chromosomal genes from bacteria and lower eukaryotes such as yeast. Complementation by yeast DNA was first shown by Struhl *et al.* [3], who cloned a 10Kb *Eco*RI fragment which complemented a non-revertible histidine auxotroph lacking the enzyme imidazole glycerol phosphate (IGP) dehydratase. In this experiment the yeast DNA, cloned in a λ phage, λgt*Schis*, was integrated into the bacterial chromosome using an integration helper phage. Transcription is probably initiated within the yeast DNA, since the λ promoters are either repressed or deleted. The same segment of yeast DNA was cloned in the Co1E1 plasmid by Ratzkin and Carbon [4], using the dA:dT tailing technique. These authors selected recombinants which would complement deletion mutants in the *leu* B gene (β iso-propyl malate dehydrogenase) or the *his* B gene (*IGP* dehydratase) of *E. coli*. The rate of reassocation of denatured DNA from the plasmid which complemented the *his* B mutation could be increased by the addition of denatured λgt*Schis* DNA, indicating that both the phage and plasmid share common sequences. The plasmid which complemented the *leu* B mutation in *E. coli* was also able to complement the equivalent gene in *S. typhimurium*.

5.1.2 *Assays for expression of novel polypeptides*
The expression of segments of eukaryotic DNA carried on plasmid vectors has been analysed in strains of *E. coli* which produce minicells (cellular buds which contain no chromosomal DNA but which do contain plasmid DNA). Mini-cells can be fractionated away from their parental *E. coli* cells and afford a means of studying the transcription and translation of plasmid genes away from host cell background. Experiments carried out by Rambach and Hogness [5], to look at the expression of cloned chromosomal DNA from *D. melanogaster* in mini-cells,

indicate that a very small proportion of this DNA can be expressed in *E. coli*. Only four recombinant plasmids, out of 37 examined, encoded novel polypeptides in their experiments. The total amount of *D. melanogaster* DNA contained within these plasmids was 400Kb, of which only 1% was both efficiently transcribed and translated within *E. coli*.

This is probably a reflection of both the absence of correct controlling signals for transcription and translation in the prokaryotic cell, and more fundamental differences in gene organisation between prokaryotes and eukaryotes. These differences and the experimental approaches which uncovered them are explored in Chapter 6. The most striking difference is that many eukaryotic genes contain intervening sequences which disrupt the continuity of the coding region. These sequences are transcribed but are subsequently spliced out of the mRNA by processing enzymes. As there is no selection pressure to maintain the coding capacity of these intervening sequences the probability is high that either they contain nonsense codons or that they disrupt the coding phase of the rest of the gene. If then, a transcript of the gene and its intervening sequences were translated in *E. coli*, it is unlikely that such a polypeptide would be functional. For these reasons it is necessary to supply the bacterium with a gene in which the intervening sequences are no longer present. We shall see later in this Chapter that this is conveniently done either by synthesising a double stranded copy of the mRNA for the eukaryotic gene or by chemically synthesising the gene.

A convenient assay for the synthesis of a foreign polypeptide is needed. The assay for novel polypeptides by SDS polyacrylamide gel electrophoresis of extracts of *E. coli* mini-cells is time-consuming and yields very little information. A number of rapid screening techniques have now been described which are similar in concept to the colony hybridisation technique of Grunstein and Hogness (see Chapter 3), but rather than enabling the detection of bacteria containing a particular nucleic acid sequence by specific hybridisation with a radiolabelled nucleic acid probe, they enable the detection of colonies which are producing particular polypeptides which will form specific complexes with antibodies. The first attempts at immunoassays detected visible immuno-precipitates around the plaques or colonies when the specific antiserum was contained in the growth medium. The sensitivity and rapidity of the screen has been greatly increased by the use of radiolabelled antibodies and a solid phase support for the antibody. Bacteria containing recombinant DNA are replica plated onto an agar dish and, following growth, the cells within the colonies can be lysed either by exposure to chloroform vapour or by thermo-induction of λCl857 prophage. The antibody, linked to solid phase support, is brought into gentle contact with the lysed cells and left for several hours in order to allow adsorption of antigen to the antibody. Other antigenic determinants on the bound antigen are then detected with suitable radiolabelled antibody.

Unreacted antibody can be washed away, and the position

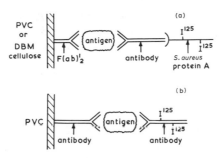

Fig. 5.1 Screening with radiolabelled antibodies.

of the complex can be determined by autoradiography, which therefore also localises on the master petri dish the position of the bacterial cells which are synthesising the antigen. In the technique of Broome and Gilbert [6] the antibody is bound to polyvinyl sheets which provide the solid phase support. The same immunoglobulin fraction, but radiolabelled with ^{125}I, is used as a probe for the antigen bound to the immobilised antibody (Fig. 5.1(b)). Ehrlich *et al.* [7] used F(ab)$^1_2$ fragments, derived by pepsin digestion of the immunoglobulin, bound to either polyvinyl sheets or diazo-benzyloxymethyl cellulose paper [8]. This is incubated first with the antigen and then with undigested antiserum, the FC portion of which will bind radiolabelled *Staphylococcus aureus* A protein (Fig. 5.2(a)).

5.1.3 *Expression of vertebrate genes in E. coli*

5.1.3.1 *Directed synthesis of fused polypeptides.* Of two major approaches used to achieve expression of vertebrate polypeptides in *E. coli*, one involves the construction of a hybrid gene with the N terminal region of a bacterial protein joined to the C terminus of a vertebrate protein. In the other approach the vertebrate gene is placed downstream from prokaryotic controlling elements within the plasmid. As an example of the first approach, I will discuss in some detail the expression of a chemically synthesised gene for the hormone somatostatin [9], a tetradecapeptide which has the physiological role of inhibiting the secretion of growth hormone, glucagon and insulin. The gene was chemically synthesised, using the triester method (see Chapter 2), as eight single stranded DNA segments which anneal in an overlapping manner to give a double stranded DNA segment having single stranded projections at each and corresponding to the cohesive ends produced by *EcoRI* and *BamHI*. Codons were chosen so as to eliminate undesirable base pairing of the eight synthetic segments and to include those codons which are probably used preferentially in *E. coli*. The synthetic gene was terminated by two nonsense codons and preceded by a methionine codon. The fused polypeptide product can be cleaved by cyanogen bromide at this methionine residue in order to convert it to the functional form of the hormone.

39

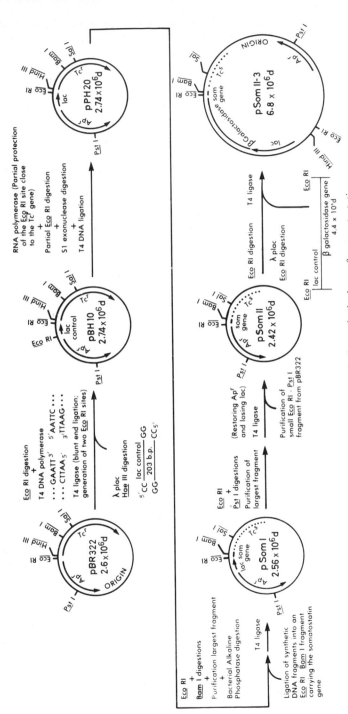

Fig. 5.2 Protocol for cloning the chemically synthesised gene for somatostatin. (from [9]; copyright 1977 American Association for the Advancement of Science).

Two plasmids were constructed in which the synthetic gene was inserted into the *E. coli* β galactosidase gene at two different sites (Fig. 5.2). In the first of these, pSoml, the synthetic gene was linked to a 203 nucleotide *Hae*III restriction fragment of the *lac* operon (isolated from a λ*plac* phage) which carries the *lac* promoter, catabolite activator protein binding site, operator, ribosome binding site and the first seven codons for the *Z* gene. This *Hae*III fragment was first joined by blunt end ligation to the *Eco*RI cleaved pBR322 which had had its *Eco*RI cohesive ends repaired by T4 DNA polymerase. This procedure regenerates *Eco*RI sites flanking the *lac* segment. One of these recognition sites was then deleted. The *Eco*RI site, close to the tetracycline resistance gene, is at an RNA polymerase binding site and can be protected with RNA polymerase from *Eco*RI digestion. The single stranded *Eco*RI cohesive termini at the other site of cleavage were removed using S1 nuclease and the plasmid, pBH20, recyclised by blunt end ligation using T4 ligase. The synthetic somatostatin gene was then inserted between the *Eco*RI site and *Bam*HI site of pBH20, using techniques discussed in Chapter 2. This recombinant plasmid was designated pSoml.

The somatostatin gene was also inserted in phase into the β galactosidase gene at an *Eco*RI site near the C terminus. This was achieved by replacing the *Eco*RI *Pst*I fragment of pSoml with the *Eco*RI *Pst*I fragment of pBR322 and subsequently inserting a large *Eco*RI fragment containing the *lac* operon control region and most of the *Z* gene into the *Eco*RI site of this new plasmid. This second recombinant was designated pSomII-3. Somatostatin can be detected immunologically in extracts of cells carrying pSoml or pSomll-3 following treatment with cyanogen bromide, although detection is easier in cells producing the large fused peptide (encoded by pSomll-3). This is thought to be due to stabilisation by the βgalactosidase moiety of the larger fused polypeptide to degradation by endogenous proteases. The generality of this approach is limited to the production of polypeptides which do not contain internal methionine residues, and it still remains to be seen whether cultures of cells carrying these recombinant plasmids can be efficiently scaled up to permit economic purification of the cyanogen bromide-generated hormone.

The chromosomal genes for many proteins from higher organisms contain intervening sequences which are removed from the biologically active mRNA. A cDNA copy of mRNA therefore represents an alternative source of polypeptide coding sequences which might be expressed in *E. coli*. The basic technique for cloning cDNA is perhaps best exemplified by the work of Maniatis *et al.* [10] in their cloning of a cDNA copy of the rabbit β globin gene (Fig. 5.3). Synthesis of the first strand is primed from oligo dT annealed to the polyA tract on the 3' terminus of the message. The RNA template is then hydrolysed and the second strand synthesis primed from hairpin structures which are formed at the 3' terminus of the first strand. The hairpin can then be cleaved with S1 endonuclease and the double

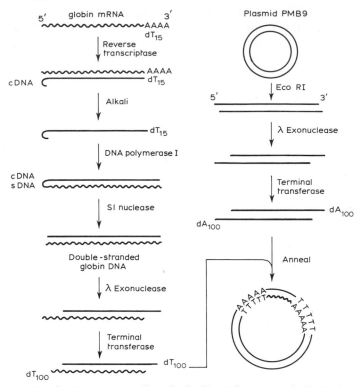

Fig. 5.3 Scheme for cloning cDNA (from [10]; copyright M.I.T.).

stranded cDNA can be cloned in a plasmid vector by the tailing technique (see Chapter 2).

Villa-Kamaroff and colleagues have isolated mRNA from tumour cells which overproduce insulin and have used this as a template for the synthesis of double stranded cDNA which they have cloned by dG:dC tailing into the *Pst*I site in the ampicillinase gene of pBR322 (see Chapter 2) [11]. Ampicillinase is a periplasmic protein synthesised with a 23 amino acid leader, which is thought to direct the protein across the cell membrane and which is removed during this process. This has enabled the detection by the solid-phase radioimmune assay [6] of cells carrying recombinant plasmids which are expressing ampicillinase-insulin hybrid polypeptides. In one such cell a hybrid polypeptide is produced in which the two domains are connected by six glycine residues encoded by the dG:dC connector region.

A slightly different approach was used by Seeberg and coworkers [12] to achieve expression of rat pre-growth hormone sequences. Chemically synthesised *Hind*III linkers were attached to cDNA to the rat pre-growth hormone mRNA which was then cloned in the *Hind*III site of pBR322. This cDNA *Hind*III fragment was transferred into pMB9 to give a plasmid (pMB9-RGH) in which tetracycline resistance is expressed at low levels. The cDNA has a unique *Pst*I site a few base pairs distal to the initiation codon for the polypeptide. It was therefore possible to replace the *Pst*I-*Bam*HI fragment from pMB9-RGH with the *Pst*I-*Bam*HI fragment of pBR322. This reconstitutes the fully active tetracycline resistance gene and fuses the N terminus of the ampicillinase gene of pBR322 with the sequences coding for the 214 C-terminal amino acids of rat pre-growth hormone. As in the case with the cloned insulin gene colonies of cells producing the fused polypeptide could be identified by the immunological screening technique [6] using radiolabelled anti-rat growth hormone immunoglobin as a probe.

The synthesis of foreign polypeptides in *E. coli* is likely to be of considerable importance in antigen production for the development of vaccines. Hepatitis B viral DNA has recently been cloned in pBR322 [13]. The virus is widespread in man and produces several types of chronic liver disorders. The limited amount of virus available from infected patients and the inability to grow the virus in cultured cells has seriously hindered its molecular characterisation and the development of vaccines. The viral DNA is a double-stranded circular molecule of approximately 3Kb and has a large single-stranded gap which must be repaired with an endogenous viral polymerase before the molecule can be subject to restriction-endonuclease digestion and molecular cloning. Three viral proteins have been recognised antigenically: a viral surface antigen, an inner viral core antigen and a third antigen, the e-antigen, which is detected in the blood of some infected individuals. The complete viral DNA sequence has now been determined from cloned molecules and knowledge of the terminal amino-acids of the surface and core antigens has enabled their genes to be localised within the complete nucleotide sequence. Many of the bacterial clones containing hepatitis viral DNA have been tested for their ability to produce viral antigens and several have already been shown to synthesise a polypeptide which is immunologically recognisable as core antigen.

5.1.3.2. *Direct expression.* It seems certain that in some cases transcripts of cDNA inserted into the *Pst*I site of pBR322 are translated, not to give fused polypeptides, but with translation being correctly initiated at the AUG codon in the cDNA. The biological expression of a DNA sequence coding for mouse dihydrofolate reductase (DHFR) is such a case. This enzyme, which catalyses the conversion of dihyrofolic acid to tetrahydrofolic acid, is overproduced in mammalian cell lines selected for their ability to grow in the presence of methotrexate. The mRNA for DHFR is readily purified from such cells, and cDNA synthesised from such a message was cloned by the dG:dC tailing technique into the *Pst*I site of pBR322 [14]. Mammalian DHFR has a lower affinity for the drug trimethoprim than does the bacterial enzyme. This permits direct selection for *E. coli* cells carrying functional recombinant plasmids, since they will grow in a higher concentration of the drug. Since *E. coli* χ1776 is *thy*$^-$ and so resistant to

43

trimethoprim, the selection was carried out in the thy^+ derivative $\chi2282$. Independent clones containing recombinant plasmids showed differing degrees of resistance to trimethoprim. In the case of one clone which showed resistance to a high concentration of the drug, the plasmid vector was shown to be interrupted at its PstI side by 11 dG residues followed by an AUG codon and then the codon for the first amino acid of DHFR, which is no longer the same translational reading frame as the β lactamase gene. Furthermore the polypeptide produced in these bacteria was identical to the mammalian enzyme in a number of antigenic tests, in its sensitivity to methotrexate and trimethoprim, and in its electrophoretic mobility. It seems therefore that in this recombinant molecule, the poly dG sequence serves as a binding site for the bacterial ribosome as postulated by Shine and Dalgano [15]. Support is given to this explanation by the finding that the enzyme is not as efficiently synthesised from other recombinant molecules in which there are differing numbers of residues between the N terminus of the gene and the dG:dC junction with the cloning vector.

cDNA copies of human leukocyte interferon mRNA have recently been cloned in the PstI site of pBR322 by dG:dC tailing [16]. It seems that such clones can also be expressed within E. coli to give non-fused proteins. This polypeptide has considerable potential as an antiviral agent and possibly also as an inhibitor of tumour growth [17], and it remains now only to increase the yields of this polypeptide in E. coli so that the potential can be investigated.

In the above cases the correct initiation of translation at the Met codon of the mammalian gene was somewhat fortuitous. Goeddel and coworkers [18], however, designed a recombinant plasmid in which the human growth hormone would be directly expressed. In this case, if translation were to be initiated at the natural Met codon in cloned cDNA, for example, then the polypeptide product would be the pre-hormone which could not of course be correctly processed by the E. coli cells. Since the mature hormone has 191 amino acids, the chemical synthesis of its gene would be too time-consuming. A gene was constructed, therefore, in which the C terminal coding region was a restriction fragment of cDNA to the pre-hormone mRNA and in which the N terminal coding region was chemically synthesised. A 550 base pair HaeIII fragment was cleaved from double stranded cDNA made against total polyadenylated RNA from human pituitaries. This fragment was gel purified and tailed with dC residues and cloned into Pst cleaved pBR322 tailed with dG residues (pHGH31 ; Fig. 5.4). This procedure regenerates both the HaeIII sites (GG↓CC) on the ends of the cDNA and the PstI site (CTGCA↓G) of pBR322 (see also Fig. 2.) and so the HaeIII fragment can be amplified by propagation in the plasmid and subsequently recovered. The HaeIII site occurs at amino acids residues 23 and 24 and so it was necessary to chemically synthesise double stranded DNA which would encode the sequence up to these residues. This was achieved using the approach described earlier for the somatostatin gene, in which the coding region

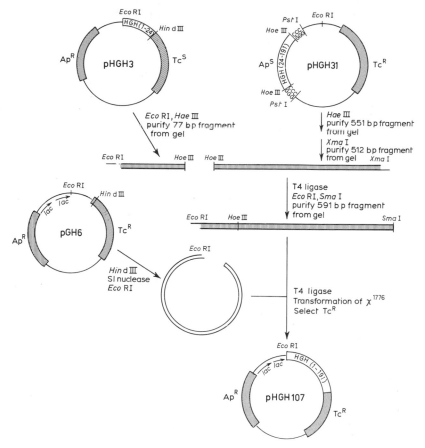

Fig. 5.4 Construction of a plasmid for the expression of human growth hormone from three cloned components (from [18]; copyright © 1979 Macmillan Journals Limited).

was preceded by a Met codon and *Eco*RI cohesive terminus. The *Hae*III sequence in the gene was followed by a *Hind*III terminus so that the overlapping fragments could be cloned between the *Eco*RI and *Hind*III sites of pBR322 (pHGH3. Fig. 5.4). The region which was to control gene expression was a 285 base pair *Eco*RI fragment containing two *lac* promoter elements. This was cloned into pBR322 and then one of the two resulting *Eco*RI sites destroyed, again using procedures described in the cloning of the somatostatin gene (pGH6; Fig. 5.4). Elements of these three recombinant plasmids were then combined as outlined in Fig. 5.4. The amplified chemically-synthesised segment was cleaved from its plasmid with *Eco*RI and *Hae*III and the purified fragment ligated to a *Hae*III/*Xma*I fragment from within the cloned cDNA (the *Xma*I site occurs 3 nucleotides distal to the termination codon of the gene). Cleavage of these ligation products with *Sma*I which recognises the same site as *Xma*I, but generates flush ends,

45

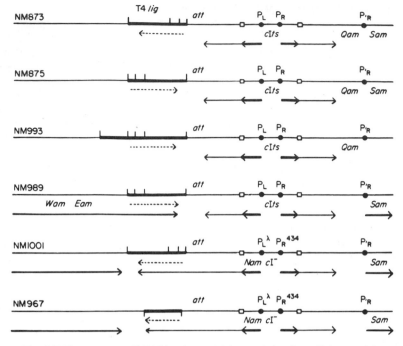

Fig. 5.5 The genomes of λ T4 Lig phages (with permission from [20]; copyright Academic Press Inc. (London) Ltd.).

produced a 591 base pair fragment with an *Eco*RI terminus and a fully base paired *Sma*I terminus. This could be inserted distally to the *lac* promoters cloned in pBR322. The *lac* promoter segment contains a ribosome-binding sequence AGGA which occurs in the final recombinant, pHGH107, 11 base pairs before the initiator codon of the human growth hormone gene. Since this sequence occurs 7 base pairs in front of the initiator codon of the *lacZ* gene, a derivative of pHGH107 was constructed in which 4 nucleotides from this region were deleted. This was achieved by cleaving the plasmid with *Eco*RI, removing the single stranded tails with S1 nuclease and religating the blunt ends with T4 ligase. Surprisingly it was found the human growth hormone was more efficiently produced in the plasmid in which there were 11 residues between the putative ribosome binding site and the Met codon. The experiment still illustrates that the effectiveness whereby the transcription products of the *in vitro* recombinant gene are translated does depend upon the sequence which immediately precedes the initiator codon. Similar experiments have been carried out in which the SV40 t antigen gene has been placed under control of a similar segment of the *lac* operon [19]. Again it was found that the level of expression of the SV40 gene depended upon the position of the initiator codon from the 'Shine-Dalgano' sequence in the *lac* DNA.

5.2 Expression from phage λ promoters

Most phage vectors accept DNA in a region of the genome where either the P_L promoter of the P_R' promoter could be used for efficient transcription (see Chapter 4 for a description of the transcription circuits of phage λ). The work of Murray et al. [20] is an ideal case study of the comparative yields of T4 DNA ligase from a series of λ-T4 recombinants constructed in vitro (T4 lig phages) in which transcription is either from T4 promoters or from the P_L' and P_R promoters. The genomes of the λ-T4 lig phages used in this study are shown in Fig. 6.1. Expression of the T4 ligase gene in phages λNM873, λNM875 and λNM993 occurs from T4 promoters when the major λ promoters are repressed. The amplification of the enzyme yield depends upon the gene dosage, which increases during the lytic interaction with the host. The lytic interaction can be initiated by thermoinduction of lysogens since the CI gene contains a temperature-sensitive mutation. The incorporation of mutations in Q and S drastically reduces late transcription and prevents cell lysis and so allows a high yield of phage DNA and retention of all phage determined polypeptides within the cell. The yield of ligase from such phage is not, however, particularly efficient.

The phage λNM1001 and λNM967 are phage in which the ligase gene should be transcribed from P_L. Ideally the phage should be cro^- to relieve the repressive effect of the cro gene product on transcription from P_L and P_R. Since cro^- phage replicate their DNA poorly, it is possible to use a hybrid immunity phage that contains P_R^{434} and cro^{434} along with the P_L of λ which is not repressed by cro^{434} gene product. Such a hybrid immunity phage can only undergo lytic interactions with the host. The most efficient amplification of T4 ligase was obtained from the induction of the λNM989 prophage in which transcription is from P_R'. Efficient late transcription from this promoter requires the Q gene product and so in this case the Q is wild type. Cell lysis and packaging are eliminated by mutations in genes S, W and E (See Chapter 4).

References

[1] Clarke, L. and Carbon, J. (1975) Proc. natn Acad. Sci. USA. **72**, 4361.

[2] Clarke, L. and Carbon, J. (1976) Cell, **9**, 91.

[3] Struhl, K., Cameron, J. R. and Davis, R. W. (1976), Proc. natn Acad. Sci. USA. **73**, 1471.

[4] Ratzkin, B. and Carbon, J. (1977) Proc. natn Acad. Sci. USA. **74**, 487.

[5] Rambach, A. and Hogness, D. S. (1978) Proc. natn Acad. Sci. USA. **74**, 5041.

[6] Broome, S. and Gilbert, W. (1978) Proc. natn Acad. Sci. USA. **75**, 2746.

[7] Erlich, H. A., Cohen, S. N. and McDevitt, H. O. (1978) Cell **13**, 681.

[8] Alwine, J. C., Kemp, D. J. and Stark, G. R. (1977) Proc. natn Acad. Sci. USA. **74**, 5350.

[9] Itakura, K., Hirose, T., Crea, R., Riggs, A. D., Heyneker, H. L., Bolivar, F. and Boyer, H. W. (1977) Science **198**, 1056.

[10] Maniatis, T., Kee, S. E., Efstratiadis, A. and Kafatos, F. (1976) *Cell* **8**, 163.

[11] Villa-Komaroff, L., Efstratiadis, A., Broome, S., Lornedico, P. Tizard, R., Naber, S. P., Chick, W. L. and Gilbert, W. (1978) *Proc. natn Acad. Sci. USA.* **75**, 3727.

[12] Seeburg, P. H., Shine, J., Martial, J. A., Ivarie, R. D., Morris, J. A., Ullrich, A., Baxter, J. D. and Goodman, H. M. (1978) *Nature* **276**, 795.

[13] Burrell, C. J., Mackay, P., Greenaway, P. J., Hofschneider, P. and Murray, K. (1979) *Nature* **279**, 43.

[14] Chang, A. C. Y., Nunberg, J. H., Kaufman, R. J., Erlich, H. A., Schimke, R. T. and Cohen, S. N. (1978) *Nature* **275**, 617.

[15] Shine, J. and Dalgano, L. (1974) *Proc. natn Acad. Sci.* **71**, 1342.

[16] Nagata, S., Taira, H., Hall, A., Johnsrud, L., Streuli, M., Ecsodi, J., Boll, W., Cantell, K. and Weissmann, C. (1980), *Nature*, **284**, 316.

[17] Stewart, W. E. (1979) 'The Interferon System' Springer Verlag, Wien and New York.

[18] Goeddel, D. V., Heyneker, H. L., Hozumi, T., Arentzen, R., Itakura, K., Yansura, D. G., Ross, K. J., Miozzari, G., Crea, R., Seeburg, P. H. (1979) *Nature* **281**, 544.

[19] Roberts, T. M., Bikel, I., Rogers Yocum, R., Livingstone, D. M., Ptashne, M. (1979) *Proc. natn Acad. Sci. USA* **76**, 5596.

[20] Murray, N. E., Bruce, S. and Murray, K. (1979). *J. mol. Biol.* **132**, 493.

6 Methods for the physical characterisation of cloned segments of chromosomal DNA from higher eukaryotes

Genetic engineering techniques provide a means to purify single genes and their associated sequences from complex genomes. The haploid genome of the mouse, for example, contains approximately 3×10^6 Kb of DNA, and so the purification and characterisation of a gene such as that for β globin was impossible before the advent of such techniques. The cloning technology not only allows the purification of the gene but also its amplification during propagation in *E. coli*. Large quantities of cloned sequences can therefore be prepared and this permits their physical characterisation. In this Chapter we will examine the organisational features of several eukaryotic genes. We will see how the information gained goes some way to solving such paradoxes as why most eukaryotes have at least an order of magnitude more DNA than is apparently required to code for their structural genes, and why gene transcripts in the nuclei of most higher organisms are very much larger and more heterogeneous than the functional mRNAs in the cell cytoplasm. Our picture of chromosome organisation and gene control in eukaryotes is, however, still very incomplete in spite of the information explosion generated by the *in vitro* recombinant DNA technology. Rather than review current knowledge on gene organisation, I will describe first the techniques which allow cloned sequences to be localised on chromosomes and then the techniques which are used to generate physical maps of the cloned DNA segments themselves. In so doing I will illustrate the basic organisational features of several eukaryotic genes.

6.1 Mapping cloned DNAs to their chromosomal origins

6.1.1. *In situ hybridisation*

This technique permits the cytological localisation of nucleotide sequences on chromosomes squashed onto glass microscope slides. The chromosomal DNA is first denatured and then incubated with either ^{3}H-DNA or ^{3}H-RNA which anneals to its complementary sequence in the chromosomes. The unhybridised material is washed away and the position of hybridisation localised by autoradiography. Unfortunately the sensitivity of this technique is such that it can only be used to detect highly repetitive sequences, such as buoyant-density satellite DNAs or clustered moderately-repetitive sequences such as ribosomal genes in the diploid chromosome complement of most higher eukaryotes. This problem is not encountered with organisms such as the dipteran flies which have polytene chromosomes. *Drosophila melanogaster*, for example, has four pairs of chromosomes which in certain tissues replicate to give ploidies of up to 1056. The chromatids do not separate but instead become laterally aggregated and form giant chromosomes. The centromeric repetitive sequences do not undergo as many rounds of replication as do the rest of the chromosomal sequences and in the polytene state they aggregate to form the heterochromatic chromocentre. The homologous arms of each chromosome pair also aggregate so that a polytene squash has the appearance of five long chromosomal arms radiating out from the chromocentre (Fig. 6.1). These arms each have a characteristic banding pattern, and there is a long-established correlation between genetic and cytological chromosome maps. The endomitosis of *D. melanogaster* chromatids to form the polytene chromosomes ensures that a gene which may be described as 'unique' within the haploid genome is present at one chromosomal location in sufficient copies to permit its detection by *in situ* hybridisation [1, 2].

It is not only single-copy sequences which hybridise to single chromosomal regions: the genes for histones [3] and the genes for 5SrRNA [4], for example, are each moderately-repetitive, tandemly-arranged gene families which hybridise *in situ* to the banding positions 39DE and 56EF respectively. The telomeres of each chromosome also contain tandemly-repeating sequences [5]. Other moderately-repetitive sequences are found dispersed around the genome and several of these families of sequences code for abundant polyadenylated RNA molecules. The first of these families to be identified was named after the cloned segment cDm412 which codes for a poly-adenylated RNA 7Kb in length [6]. Radiolabelled cDm412 sequences were found to hybridise *in situ* to about 70 euchromatic regions. Restriction fragments from within

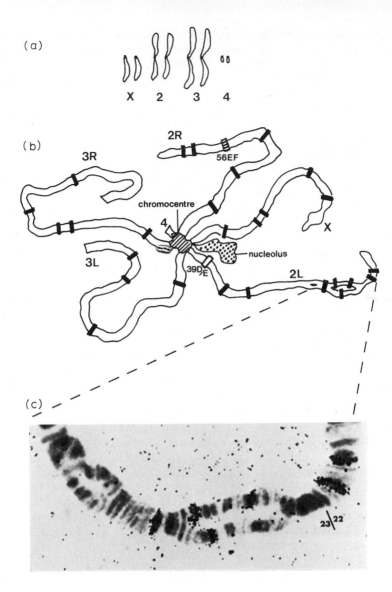

(a)

X   2   3   4

(b)

2R

3R

56EF

chromocentre

4

nucleolus

3L

39D/E

2L

X

(c)

23 22

Fig. 6.1 Diagrammatic representation of mitotic diploid chromosomes (a) and polytene chromosomes (b) of *D. melanogaster*. The
in situ hybridisation sites of some cloned genes are indicated by
boxes over regions of the polytene chromosome. The filled boxes
represent the Dm412 sequence; the open box at 39D/E, the histone
genes; the hatched box at 56EF, the 5S rRNA genes; the nucleolus
is the site of hybridisation of cloned rDNA and many moderately
repetitive sequences hybridise to the chromocentre. The micrograph
(c) shows one localised area of hybridisation of 412 sequences
(see text for details) (from [8]; copyright M.I.T.).

the coding sequence show hybridisation to 33 of these sites. cDm412 therefore contains two types of moderately-repetitive DNA, a non-coding sequence present at about 40 chromosomal sites, and a coding sequence present at 33 different sites. The Dm412 coding sequence is not, however, found at these sites in all D. *melanogaster* strains [7, 8]. Figure 6.1 shows asynapsed second chromosome homologues from an F1 Seto × Oregon R individual in which none of the hybridising sites in the asynapsed region are common between the two parental strains. The distribution of restriction fragments homologous to Dm412 and similar gene families within the chromosomal DNA of tissue culture cell lines suggests that the genes are mobile and can transpose around the genome in a manner analogous to the prokaryotic transposons. Like prokaryotic transponsons they have a terminally-redundant sequence which for 412 elements is 0.5Kb in length. The terminally-redundant sequence of another family, called copia, shows weak homology to one of its insertion sites, the tandemly repeated sequence found in the telomeres. This suggests a model for trans-position whereby copia genes (or genes from the other dispersed families) could cyclise through their redundant sequence to become episomal and subsequently re-integrate into the genome by homo-logous recombination [8].

There are probably several such gene families which together account for several per cent of the D. *melanogaster* genome and it is likely that they could correspond to the translocatable mutator genes which have been detected genetically [9]. Sequences of this kind could be a general feature within eukaryotic genomes. DNA sequences with a similar structure have recently been described in the yeast genome [10] and the organisational features of both these and the comparable *Drosophila* genes show great similarities to the inte-grated DNA genomes complementary to the retrovirus RNAs of higher eukaryotes.

6.1.2 *Somatic cell hybrids*

Cloned genes have been assigned to mammalian chromosomes using the nucleic acid hybridisation techniques which we will discuss in Section 6.2.2, in combination with well established techniques of somatic cell genetics. Interspecific cell hybrids can be readily generated by sendai virus or polyethylene glycol-mediated fusion, but there is a tendency for chromosomes from one of the parental cell types to be lost from the hybrid. It is possible to take hybrid cell lines in which a small number of chromosomes from this parent are stably retained and analyse their DNA for sequences complementary to a cloned gene. In this way the genes for human γ, δ and β globin, for example, were localised to the short arm of chromosome 11 [11, 12]. Although this technique does not have the precision of *in situ* mapping onto polytene chromosomes, it is nevertheless still possible to make use of available chromosome translocations in order to localise the cloned gene to a broad region of a chromosome.

6.2 Electrophoretic mapping techniques

6.2.1 *Restriction mapping*

Restriction endonuclease cleavage sites can usually be mapped within plasmid or phage DNA by the electrophoretic analysis of fragments produced by partial digestion with a single restriction endonuclease, or by complete digestion with a mixture of restriction endonucleases. Consider the plasmid pBR322 (Fig. 3.3) for example; if this DNA is cleaved with *Eco*RI, a linear molecule is generated whose size can be determined to be 4.36 Kb. If the DNA is cleaved with both *Sal*I and *Eco*RI, then fragments of 0.38Kb and 3.98Kb are generated. This positions the *Sal*I cleavage site 380 nucleotide pairs on one or other side of the *Eco*RI site. These two alternatives can then be discriminated by carrying out double digestions with *Eco*RI or *Sal*I and some other restriction endonuclease, for example *Pst*I. In this way an unambiguous cleavage map can be constructed. If sites for one enzyme are clustered together without any internal reference cleavage site for a second enzyme then it is usually possible to order the clustered sites from the size of fragments generated by partial restriction cleavage. This problem often arises with enzymes which recognise tetranucleotide sequences and which therefore cleave DNA at frequent intervals.

A refinement of this partial-digestion technique has been applied to the analysis of repetitive sequence elements in the non-transcribed spacers of tandemly repeating rDNA units. The technique may be illustrated from the work of Boseley *et al.* [13] on a cloned *Eco*RI fragment containing the non transcribed spacer (NTS) and adjacent 28S and 18S sequences of *Xenopus laevis* rDNA. The *Eco*RI fragment was cleaved with *Bam*HI, and after dephosphorylation using bacterial alkaline phosphatase, was labelled at its 5′ ends by rephosphorylation with radiolabelled phosphate from γ-^{32}P-ATP using T4 polynucleotide kinase. The terminally labelled *Eco*RI-*Bam*HI fragments contain asymmetric cleavage sites for the enzymes *Hin*dIII, *Hha*I and *Hinf*1, and so cleavage with either of these enzymes generates easily separable left and right 'halves', each of which will then have a terminal label at only one of its 5′ ends (Fig. 6.2, Upper Panel). Partial digestion of these 'half molecules' with an endonuclease for which there are multiple sites will generate an overlapping series of molecules which can be fractionated by gel electrophoresis. A subset of these fragments, which have a common labelled terminus, can be visualised by autoradiography (Fig. 6.2, middle panel). The order of these labelled fragments and their molecular lengths corresponds directly to the order and location of the restriction sites along the molecule.

The detailed cleavage map of the non-transcribed spacer reveals three repetitious regions: region 1 has a repeating unit of 100 base pairs. It is separated from region 2 by a non repetitive element in which is found a *Bam*HI site, and which has therefore been termed a '*Bam* island'. A similar non-repetitive element separates regions 2 and 3. These two regions are virtually identical, having alternating 81/60

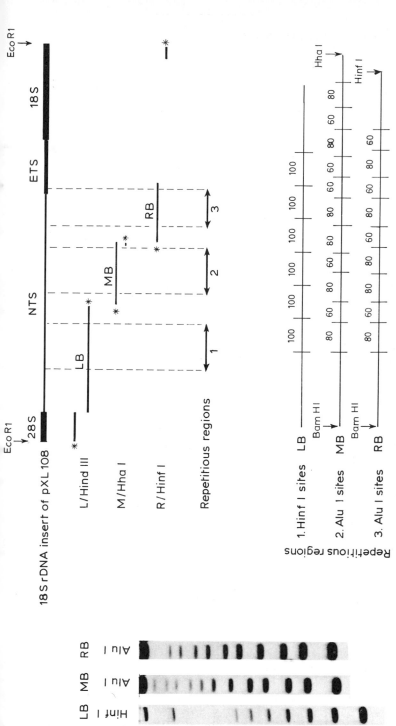

Fig. 6.2 Partial digestion mapping the *Xenopus laevis* rDNA spacer. The upper panel shows the derivation of the end-labelled fragments LB, MB and RB which are shown following partial restriction and electrophoresis in the autoradiograms on the left. The lower part of this figure shows the physical maps of the repetitive regions 1, 2 and 3 which are contained within fragments LB, MB, and RB respectively (from [13]; copyright M.I.T.).

53

base-pair arrangements, with region 3 differing from region 2 in having *Sma*I sites in the 81 base pair unit. DNA sequencing of the units in regions 2 and 3 shows that the 60 and 81 base-pair arrangements are identical, excepting a run of 21 nucleotides. There is also some homology between the nucleotide sequence of the '*Bam* island' and repeating units from both region 1 and regions 2 and 3, and between the '*Bam* island' sequence and the putative promoter region to the 5' side of the junction between the non-transcribed spacer and the external transcribed spacer (ETS). A hypothetical scheme can be constructed from these data to account for the evolution of the non-transcribed spacer by a series of reduplication events upon an ancestral sequence postulated to be at the NTS/ETS boundary. This leads to the interesting idea that, by duplicating what is potentially a promotor region, a DNA sequence has been generated which has the capability of binding many RNA polymerase molecules.

6.2.2. Gel transfer hybridisation

A restriction endonuclease which recognises a hexanucleotide sequence will cleave a random nucleotide sequence once every 4,096 (4^6) base pairs. Wild-type phage λ DNA (49Kb), for example, has 5 cleavage sites for *Eco*RI and 6 sites for *Hind*III, and so the cleavage sites in such a DNA molecule can be readily mapped by direct electrophoretic analysis of the cleavage products following ethidium bromide staining. Mammalian genomes, however, will be cleaved into approximately 10^6 specific fragments by either of these enzymes, resulting in a smear of unresolved fragments upon electrophoretic fractionation. A method has been developed by Southern [14] which permits the detection of specific classes of sequence within this smear and consequently the restriction mapping of those sequences. Following electrophoretic fractionation, the fragments of DNA are denatured within the gel and then 'blotted' onto a sheet of nitrocellulose to which they bind retaining the same relative positions that they occupied within the gel. The nitrocellulose filter is then incubated together with a ^{32}P-labelled sequence probe, which will hybridise to its complement within the smear of restricted DNA. Unhybridised probe can then be washed away and the position of hybridisation on the nitrocellulose detected by autoradiography. Autoradiographic techniques [15] can detect as little as 1 d. $\min^{-1}$ of ^{32}P and since DNA can be labelled *in vitro* to greater than 10^8 d. $\min^{-1} \mu g^{-1}$ by nick-translation [16] then it is clearly possible to use this technique to detect unique restriction fragments which represent approximately 1 part in 10^6 of a digest of mammalian DNA. This technique has found widespread application for the analysis of genomic sequences wherever cloned segments are available for use as radiolabelled probes.

The power of the technique is illustrated by the studies of Jeffreys and Flavell [17] on the structure of the rabbit β-globin gene in chromosomal DNA. These workers used an almost complete cDNA copy of the rabbit β-globin mRNA [18] as a hybridisation probe. A detailed

54

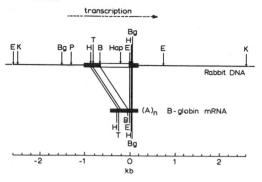

Fig. 6.3 Physical map of the rabbit β-globin gene and surrounding chromosomal DNA in comparison with the cleavage sites predicted from the mRNA sequence (from [17]; copyright M.I.T.)

restriction analysis of the chromosomal β globin sequences showed them not to be present in a continuous stretch of DNA, but instead to be in at least two pieces separated by a 600 base pair non-β-globin sequence. This conclusion was based on the mapping of intragenic sites for BamHI, TaqYI and HaeIII 600 base pairs too far to the left of an intragenic EcoRI site by comparison with the map of the cDNA clone, and also on the fact that the enzyme HapII cuts the chromosomal gene in half whereas there is no HapII site within the cDNA (Figure 6.3). Such discontinuity of eukaryotic genes, first recognised in the rDNA of D. melanogaster [19, 20], is now recognised as a common organisational feature. Globin genes from rabbit, mouse and human have now been cloned and each shown to have two intervening sequences, a larger one corresponding to the sequence described by Jeffreys and Flavell, and a smaller one which in the case of a mouse gene is 116 bases long [21]. Flavell and his co-workers have continued to use Southern's blotting technique to analyse the organisation of the human globin genes. In the foetus the major form of haemoglobin is HbF (α_2 γ_2). The γ chains are encoded by two non-allelic genes whose expression ceases at birth as HbF is replaced by HbA ($\alpha_2\beta_2$) and a low level of HbA$_2$ ($\alpha_2\delta_2$). The cloned human β globin cDNA [21] has been used as a hybridisation probe for both the β gene and the closely related δ gene in chromosomal DNA. The two genes can be distinguished since, by increasing the stringency of the hybridisation washes, it is possible to melt out those probe sequences hybridised to the δ gene. These experiments showed that both the β and δ genes contained intervening sequences and that the two genes were closely linked. Furthermore, these genes were fused in the DNA of a patient with haemoglobin Lepore, where the β-like polypeptide chain consists of an N-terminal δ-sequence fused to the C-terminal β-sequence. The restriction map of the Lepore gene is consistent with the gene fusion having occurred by unequal crossing over between the closely related genes [22]. The availability of cloned cDNAs to probe

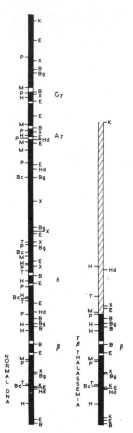

Fig. 6.4 Linkage of the human, γ, δ and β globin genes (redrawn from [26]; copyright © 1980 Macmillan Journals Limited).

for the foetal genes (Gγ and Aγ) led to the establishment of the physical linkage map shown in Fig. 6.4 [23, 24].

Examination of patients exhibiting a hereditary disease in which expression of the foetal genes persists shows that the entire δ-β region is deleted. There are several other inherited diseases known as the thalassemias in which the level of expression of the globin genes is altered. The molecular basis for several of these diseases is not understood, but it seems from restriction mapping data that in some cases the genetic lesion is a sequence deletion. In a case of $\delta\beta^\circ$ for example, part of the δ gene and the entire β gene is deleted [25]. Flavell and his colleagues have recently examined patients with a rare δ-β thalassemia, in which severe anemia is seen in new-borns and later develops into a β thalassemia [26]. They find that the Gγ, Aγ and δ globin genes have been deleted, but that the endpoint of the deletion is 2.5Kb to the 5' of the β globin gene which is still intact (Fig. 6.4). It is possible that the inactivity of this β globin gene results from a second mutation too small to be detected in this type of study, but it is equally possible that the deletion has removed a controlling element that could act either in *cis* or *trans* from a distant site.

6.2.3 *Mapping transcripts*

A second type of gel transfer hybridisation has been developed by Alwine *et al.* [27] in order to analyse the distribution of specific RNA sequences within a population of electrophoretically fractionated RNA molecules. The experimental approach is identical to Southern's technique, but a different solid phase support is used to bind and immobilise the RNA in the position to which it had migrated on the gel. The support is paper derivatised with diazobenzyloxymethyl groups to which single-stranded nucleic acid will bind covalently. Cloned DNA segments can be used as hybridisation probes to detect the transcripts of specific genes. This approach has found application in the analysis of the transcription products of eukaryotic genes containing intervening sequences such as the chicken ovalbumin gene. This gene for the major egg white protein, which is expressed in the chick oviduct under stimulation by oestrogen, contains seven intervening sequences. As seems to be the case for most genes which contain intervening sequences, the entire chromosomal sequence is transcribed and the intervening sequences are subsequently removed

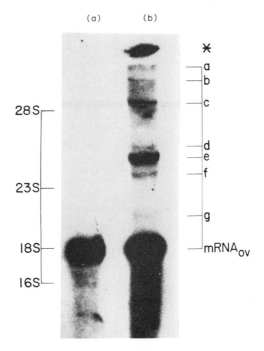

Fig. 6.5 Electrophoretically fractionated cytoplasmic (a) and nuclear (b) RNA from chick oviduct cells hybridised with the cloned ovalbumin gene (from [28]; copyright M.I.T.).

57

from the primary transcript. The coding regions of RNA are then spliced together to give the mature message. Panel A of Fig. 6.5 shows the cytoplasmic 2Kb ovalbumin mRNA detected by the technique of Alwine *et al.* The ovalbumin gene sequences are detected in seven additional bands in nuclear RNA which range up to greater than four times the length of the mRNA [28]. These RNAs represent processing intermediates from which the intervening sequences are successively removed.

Southern's technique can also be used to identify restriction fragments which encode mRNAs, but neither his approach nor the technique of Alwine *et al.* [27] can accurately position the endpoints of an RNA coding region relative to a restriction endonuclease cleavage site. It can be achieved using an approach developed by Berk and Sharp [29] for mapping viral transcripts. The technique rests upon the observation that in high concentrations of formamide DNA-RNA duplexes are more stable than the equivalent DNA-DNA duplex [30]. It is possible to thermally denature DNA duplex under such conditions, and then lower the temperature to a point at which DNA and RNA will anneal and yet which will not allow DNA reassociation. DNA which remains single-stranded can then be digested with S1 nuclease and the size of the resulting DNA-RNA duplex determined by electrophoresis, on an agarose gel. If the DNA has first been cleaved with a restriction endonuclease that cleaves within the coding region, then the length of DNA protected from S1 digestion by RNA will correspond to the distance between the end of the RNA coding region and the restriction endonuclease cleavage site [31, 32].

6.3 Mapping cloned DNAs by electron microscopy
In order to visualise DNA in the electron microscope (E.M.), a basic protein, usually cytochrome c, is added to the DNA solution and the

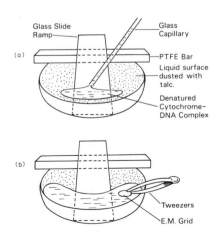

Fig. 6.6 Spreading DNA for electron microscopy.

mixture allowed to run down a glass ramp and form a molecular film over a liquid surface (the hypophase). The nucleic acid–cytochrome c complex is then picked up onto the surface of an electron microscope grid coated with a film of parlodion (a form of nitrocellulose). This is done simply by touching the coated grid onto the molecular film formed by the DNA-cytochrome solution (Fig. 6.6). The grids are stained in an ethanolic solution of uranyl acetate and subsequently subjected to low angle rotary shadowing with a platinium-palladium mixture. In general, two types of spreading technique are used. The aqueous technique (in which the hypophase and the solution containing the DNA are of ammonium acetate) allows the visualisation of double stranded nucleic acids only, and single stranded nucleic acids collapse into a bush. In another technique both single stranded and double stranded nucleic acids can be visualised when the DNA is spread over buffered formamide solutions.

6.3.1 *Denaturation mapping*
This technique permits the mapping of AT-rich regions within a DNA molecule. Such regions are the first to melt when double-stranded DNA is exposed to denaturing conditions. A partially denatured DNA molecule can be visualised by spreading at very high formamide concentrations. Alternatively the molecule can be partially denatured with alkali, and the single stranded regions fixed with formaldehyde prior to spreading in a moderate concentration of formamide such as is normally used to prevent the collapse of single stranded regions. The example shown in Figure 6.7 is a cloned segment of *Drosophila melanogaster* DNA which contains 32 tandem repeats of the gene for 5S rRNA cloned in the plasmid ColEl [33]. The plasmid vector sequences to the left of the micrograph contain a long stretch of GC

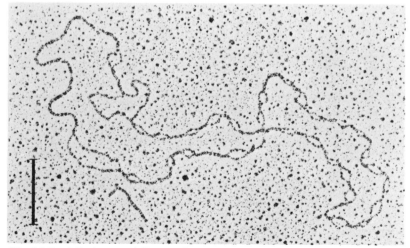

Fig. 6.7 Partially denatured 5S genes from *D. melanogaster* cloned in ColEl; Bar-1Kb (from [33]; copyright M.I.T.).

59

rich DNA, on each side of which are two large denaturation bubbles. To the right of the molecule the tandemly repeating 5S DNA forms a regular array of double-stranded sequences corresponding to the GC rich gene, and small denaturation bubbles corresponding to the AT rich spacer. The 5S rRNA genes of *Xenopus laevis* have a similar sequence arrangement, but in this case sequence analysis has identified, in addition to the 121 residues within the GC rich region which encode 5S rRNA, a highly homologous sequence, the 'pseudogene', which is 101 residues in length [34]. The function of the 'pseudogene' is still an enigma.

6.3.2 *Heteroduplex mapping*
This is probably the most widely applied E.M. mapping technique permitting the localisation of regions of sequence homology between DNA molecules. The two types of DNA molecule are mixed and denatured in alkali, and the solution is neutralised and formamide added to allow reassociation at room temperature. Following reannealing for such a length of time as to permit about 50% of homologous sequences to form duplex structures, the mixture is diluted and spread.

Heteroduplex mapping is used in the example in Fig. 6.8 to localise the regions of homology between the type I insertion (an intervening sequence) found in a high proportion of *D. melanogaster* 28S rRNA genes and homologous sequences from other heterochromatic chromosomal regions [35]. The two molecules which are heteroduplexed have each been sub-cloned from larger cloned segments of chromosomal DNA. The plasmid ckDm103B consists of most of the type I rDNA insertion cloned within the kanamycin resistance gene of the plasmid pML2. The *D. melanogaster* DNA is flanked by the inverted sequence repeat of the prokaryotic transposon (see Chapter 3) which, upon denaturation and brief reannealing of the *Eco*RI cleaved plasmid, snaps back to form a hairpin structure. In this figure, ckDm103B is annealed with *Eco*RI-cleaved pB7, a repeating unit from a tandem array of type I sequences sub-cloned into pBR322. The segments of *D. melanogaster* DNA in these two plasmids form one region of duplex ($\gamma + \varepsilon$) in which there is a single-stranded deletion loop, corresponding to sequences (δ) which are present in the rDNA insertion but absent from this unit of the tandem array. Measurements of a number of such micrographs permit the accurate mapping of the position of this deletion loop within the regions of homology, as is shown in the map of the plasmids from which these two sub-clones are derived. About 50% of these type I sequences are found in 28S genes, the remainder being in other chromosomal arrangements. The two groups of sequences are undergoing divergence, as evidenced by the variation in restriction endonuclease cleavage maps and the above heteroduplex studies, but the function of these sequences remains a mystery.

6.3.3 *Mapping regions homologous to RNA*
A number of techniques have been developed to look at RNA annealed

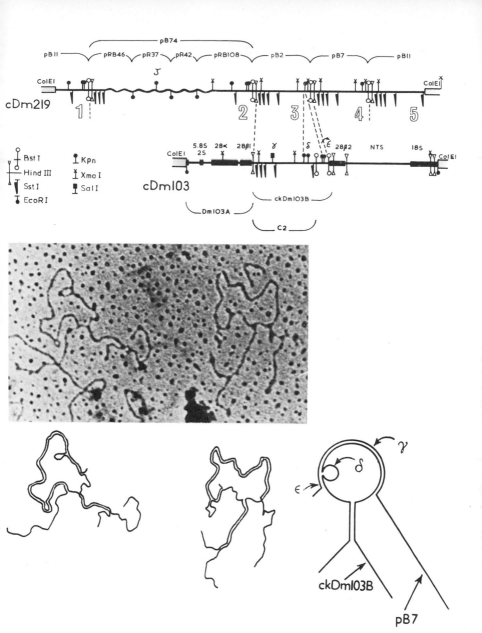

Fig. 6.8 Heteroduplex structures formed between the type I sequence insertion of *D. melanogaster* rDNA (sub-cloned in the plasmid ckDm103B) and one unit (sub-cloned in the plasmid pB7) from a tandem array of similar sequences which are known to hybridise *in situ* to the chromocenter of polytene chromosomes. The physical maps are of a cloned segment of such a tandem array (cDm219) and a unit of rDNA which contains a type I insertion (cDm103). The dashed lines between the two maps show the regions of homology between the rDNA insertion and pB2, which gives identical heteroduplex structures to the ones illustrated for pB7 (from [35]; copyright M.I.T.).

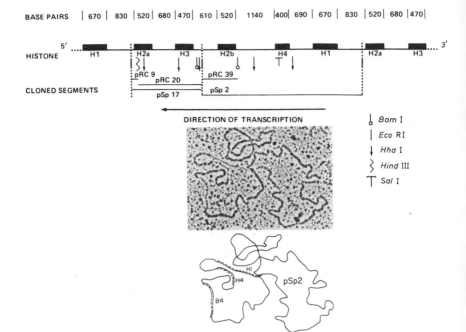

Fig. 6.9 Sea urchin histone genes. The physical map shows part of the repeating gene unit of *Strongylocentrotus purpuratus*. The micrograph is of single-stranded pSp2 DNA annealed to mRNAs for H1, H2b (labelled B4) and H4, and examined by the gene 32 protein-ethidium bromide spreading technique (from [37]; copyright M.I.T.).

to DNA. The most straightforward is simply to anneal denatured DNA with RNA and then examine the structure following formamide spreading. The interpretation of such DNA-RNA duplexes can be very difficult if one has annealed a linear DNA molecule, containing a gene plus flanking sequences, to an RNA molecule perfectly colinear with the gene, since it is visually difficult to position the junction between single-stranded DNA and DNA-RNA duplex. This is less of a problem with a eukaryotic gene containing intervening sequences which will form loops of single stranded DNA at the splice positions on the mature RNA molecule. Indeed the technique has been extensively used by Chambon and his co-workers to map the positions of intervening sequences on several genes from the chicken [36].

There are two other approaches which are extensively used to map the endpoints of genes which do not have intervening sequences. Fig. 6.9 is a micrograph of messenger RNA for three of the five sea urchin histones annealed to a segment of the histone genes and spread in the presence of T4 gene 32 protein and ethidium bromide [37]. The gene 32 protein binds specifically to single-stranded DNA and as a consequence the single-stranded regions appear fatter than double-stranded regions. The addition of the intercalating dye, ethidium bromide, to spreading mixtures gives double-stranded DNA a smoother contour. The five histone genes separated by spacer sequences are arranged as tandemly repeating units. Similar tandem arrangements are found in other organisms, although the order and relative orientation of the five genes within the repeating unit are not necessarily identical to that found in the sea urchin [3].

The second approach is that of R-loop mapping [20], which also makes use of the increased stability of DNA-RNA duplexes in high formamide concentrations. Unlike the S1 mapping technique discussed earlier in this chapter, the annealing is carried out with RNA and double-stranded DNA at a temperature which is ideally 1°C below that required for complete DNA-DNA strand separation. Under these conditions, the RNA anneals with its complementary sequence within the DNA molecule and displaces a single strand of DNA to give the R-loop.

The mapping of the coding regions upon cloned segments of mouse $\lambda 1$ immunoglobin genes has been carried out by R-loop mapping [38]. Immunoglobin peptides have a variable (V) amino-terminal region and a constant (C) carboxy-terminal region. The characterisation of cloned segments of these genes, and the application of gel transfer hybridisation studies on total genomic DNA with the cloned segments as probes, has shown that the DNA coding for the V and C regions is widely separated in embryonic cells but is brought together by a specific recombinational event in differentiated cells which produce immunoglobulin. In embryonic DNA the major part of the V region (residues 1–97) is on one DNA segment and the remainder, termed the J region (coding for residues 98–109), is on a separate DNA segment. Clones containing part of the V region (Ig13 and Ig99) give simple R loop structures (Fig. 6.10) consisting of a single loop in the region of homology between DNA and mRNA, with the 3′ end of the RNA molecule hanging free. The J segment is closely linked to the C region of the gene, but the two are separated by 1.2Kb intervening sequence. The R loop structure formed by this cloned segment (Ig25) shows two loops which correspond to DNA/RNA hybrid and displaced single stranded DNA, between which is a loop of double-stranded DNA corresponding to the intervening sequence. At the end of the structure which has the smaller R-loop is a tail of the unannealed 5′ end of the mRNA, and at the 3′ end of the gene a whisker of the polyA sequences. In DNA from the differentiated cells the 3′ end of the V region is linked to the 5′ end of the J region, and a cloned segment of this DNA forms two R loops representative of the full length of the mRNA but still retaining the intervening sequences between the J and C regions. These sequence relationships (shown diagrammatically in Figure 6.10) have been confirmed in heteroduplex experiments between the cloned DNAs.

More recently the structure of DNA coding for the mouse κ chains

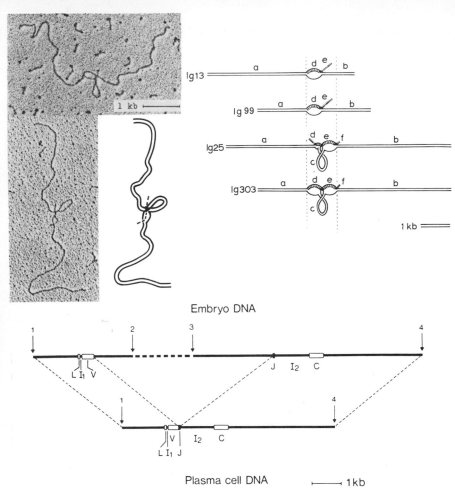

Fig. 6.10 R Loop maps of cloned λ 1 immunoglobulin genes from mouse. The micrograph is of the cloned Ig25 (see text). The R loop structures formed by four cloned segments are illustrated alongside the micrograph. The lower panel shows the arrangement of this gene in embryonic and committed tissue (from [38]; copyright M.I.T.).

has been established. In embryonic DNA the V, J and Cκ regions also exist as separate sequences, the V and J regions being brought together by recombination during differentiation. In this case, however, there are multiple copies of both the V and J regions [39, 40]. The J region sequences are all clustered to the 5′ end of the C region, and recombination can occur between any of the V and J regions. Depending upon which J region they contain, the mRNAs isolated from different myeloma cell lines will give different R loop structures with the cloned segment (Ig 146 κ) which contains the C region and five J regions. Sequencing studies suggest that the specific recombination which results in the joining of V and J regions could involve an intermediate in which the 5′ flanking regions of J segments form an inverted stem structure with the 3′ non-coding region of embryonic V genes. Antibody diversity could then in part be generated by such recombinational events.

References

[1] Wensink, P. C., Finnegan, D. J., Donelson, J. E. and Hogness, D. S. (1974) *Cell* **3**, 315.

[2] Glover, D. M., White, R. L., Finnegan, D. J. and Hogness, D. S. (1975) *Cell* **5**, 149.

[3] Lifton, R. P., Goldberg, M. L., Karp, R. W. and Hogness, D. S. (1977) *Cold Spring Harbor Symp. quant. Biol.* **42**, 1047.

[4] Artavanis-Tsakonas, S., Schedl, P., Tschudi, C., Pirrotta, V., Steward, R. and Gehring, W. (1977) *Cell* **12**, 1057.

[5] Rubin, G. M. (1977) *Cold Spring Harbor Symp. quant. Biol.* **42**, 1041.

[6] Finnegan, D. J., Rubin, G. M., Young, M. W. and Hogness, D. S. (1977) *Cold Spring Harbor Symp. quant. Biol.* **42**, 1053.

[7] Potter, S. S., Brorein, W. J., Dunsmuir, P. and Rubin, G. M. (1979) *Cell* **17**, 415.

[8] Strobel, E., Dunsmuir, P. and Rubin, G. M. (1979) *Cell* **17**, 429.

[9] Gehring, W. and Paro, R. (1980), *Cell*, **19**, 897.

[10] Cameron, J. R., Loh, E. Y. and Davis, R. W. (1979) *Cell* **16**, 739.

[11] Jeffreys, A., Craig, I. N. and Francke, U. (1979) *Nature* **281**, 606.

[12] Sanders-Haigh, L., Anderson, W. F. and Franckell (1980). *Nature* **283**, 683.

[13] Boseley, P., Moss, T., Machler, M., Portmann, R. and Birnstiel, M. (1979) *Cell* **17**, 19.

[14] Southern, E. M. (1975) *J. mol. Biol.* **98**, 503.

[15] Laskey, R. A. and Mills, A. D. (1977) *FEBS Letters* 314.

[16] Rigby P. W. J., Dieckmann, M., Rhodes, C. and Berg, P. (1977) *J. mol. Biol.* **113**, 237.

[17] Jeffreys, A. J. and Flavell, R. A. (1977) *Cell* **12**, 1097–1108.

[18] Maniatis, T., Kee, S. E., Efstratiadis, A. and Kafatos, F. (1976) *Cell* **8**, 163.

[19] Glover, D. M. and Hogness, D. S. (1977) *Cell* **10**, 167.

[20] White, R. L. and Hogness, D. S. (1977) *Cell* **10**, 177.

[21] Konkel, D. A., Tilghman, S. M. and Leder, P. (1978) *Cell* **15**, 1125.

[22] Flavell, R. A., Kooter, J. M. de Boer, E., Little, P. F. R. and Williamson, R. (1978) *Cell* **15**, 25.

[23] Little, P. F. R., Flavell, R. A., Kooter, J. M., Annison, G. and Williamson, R. (1979) *Nature* **278**, 227.

[24] Bernards, R., Little, P. F. R., Annison, G., Williamson, R. and Flavell, R. A. (1979) *Proc. natn Acad. Sci. USA.* **76**, 4827.

[25] Bernards, R., Kooter, J. M. and Flavell, R. A. (1979) *Gene* **6**, 265.

[26] Van der Ploeg, L. H. T., Konings, A., Oort, M., Roos, D., Bernini, L. and Flavell, R.A. (1980) *Nature* **283**, 637.

[27] Alwine, J. C., Kemp, D. J. and Stark, G. R. (1977). *Proc. natn. Acad. Sci. USA.* **74**, 5350.

[28] Roop, D. R., Nordstrom, J. L., Tsai, S. Y., Tsai, M-J. and O'Malley B. W. (1978) *Cell* **15**, 671.

[29] Berk, A. J. and Sharp, P. A. (1977), *Cell*, **12**, 721.

[30] Birnstiel, M. L., Sells, B. M. and Purdom, I. F. (1972) *J. mol. Biol.* **63**, 21.

[31] Tsujimoto, Y. and Suzuki, Y. (1979) *Cell* **16**, 425.

[32] Din, N., Engberg, J., Kaffenberger, W. and Echert, W. A. (1979) *Cell* **18**, 525.

[33] Davis–Hershey, N., Conrad, S., Sodja, A., Yen, P. H., Cohen, M., Davidson, N., Ilgen, C. and Carbon, J. (1977) *Cell* **11**, 585.

[34] Jacq, C., Miller, J. R. and Brownlee, G. G. (1977) *Cell* **12**, 109.

[35] Kidd, S. J. and Glover, D. M. (1980) *Cell* **19**, 103.

[36] Garapin, A. C., Cami, B., Roskam, W., Kourilsky, P., Le Pennec, J. P., Perrin, F., Gerlinger, P., Cochet, M. and Chambon, P. (1978) *Cell* **14**, 629.

[37] Wu, M., Holmes, D. S., Davidson, N., Cohn, R. H. and Kedes, L. H. (1976) *Cell* **9**, 163.

[38] Brack, C., Hirama, M., Lenhard-Schuller, R., Tonegawa, S. (1978) *Cell* **15**, 1.

[39] Sakamo, H., Huppi, K., Heinrich, G. and Tonegawa, S. (1979), *Nature* **280**, 288.

[40] Siedman, J. G., Max, E. E. and Leder, P. (1979), *Nature* **280**, 370.

7 Approaches for studying expression in eukaryotic systems

DNA sequencing of cloned eukaryotic genes has revealed several interesting common features: a potential eukaryotic promoter sequence TATAAA [1], found approximately 30 residues upstream from eukaryotic genes transcribed by RNA polymerase II; preferred bases at the boundaries between the coding parts of genes and their intervening sequences, which are postulated to be recognition sites for splicing enzymes [2]; regions of sequence homology which could present sites important in recombinational events [3] (see, for example, the discussion of immunoglobulin genes in Chapter 6). In none of these cases is there yet any proof of the functional importance of the precise nucleotide sequences. A general approach to solve these problems is to generate mutations *in vitro* at specific sites in cloned DNA segments and then to study the effects of the mutation when the DNA segments are reintroduced into eukaryotic cells or an *in vitro* system for eukaryotic gene expression. Since such an approach is in its infancy, this chapter will serve simply to draw attention first to techniques for *in vitro* mutagenesis, and second to some eukaryotic systems which are available to study the expression of cloned eukaryotic genes.

66

7.1 In vitro mutagenesis

Several means whereby restriction fragments can be rearranged within recombinant DNA molecules have been encountered in Chapter 5. Deletions can be created in a number of ways: Treatment of EcoRI-generated fragments with S1 endonuclease will remove the single-stranded termini, creating a blunt ended' molecule which can be joined to other 'blunt ended' DNA with T4 ligase, so effecting a deletion of four nucleotide pairs at the junction. Alternatively, restriction fragments can be briefly treated with either λ exonuclease, or exonuclease III followed by S1 endonuclease, to degrade the exposed single-stranded termini and so make more extensive deletions around the restriction endonuclease cleavage site. In principle, the exonuclease treatment could be carried out in a highly controlled manner if T4 DNA polymerase were used in the presence of a single deoxynucleoside triphosphate. Under these conditions, the $3' \rightarrow 5'$ exonuclease activity of T4 DNA polymerase will remove the $3'$ terminal residues of double-stranded DNA until the complement of the single deoxynucleoside triphosphate is exposed, whereupon the polymerising activity will covalently join this residue onto the $3'$OH terminus. A stable equilibrium is set up at this point, with a precise number of residues having been removed. The single stranded $5'$ termini could then be removed with S1 endonuclease.

Carbon et al, [4] have used another method to introduce a small deletion at a restriction endonuclease cleavage site. The restricted

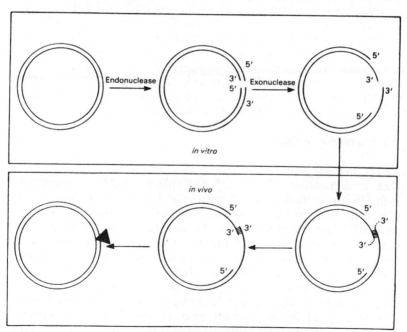

Fig. 7.1 Construction of deletion mutants by the procedure of Carbon et al. [4].

DNA is briefly treated with λ-exonuclease to generate protruding single stranded 3' termini. These molecules can then be introduced into a cell, whereupon they cyclise, presumably as a result of fortuitous homology between the single stranded termini. Molecules of this kind are evidently repaired *in vivo* by either mammalian [4] or bacterial cells [5], resulting in the generation of a deletion at the restriction endonuclease cleavage site (Figure 7.1).

Shortle and Nathans [6] have developed a method which effects a G:C→ A:T transition in the vicinity of a restriction endonuclease cleavage site. Supercoiled DNA is nicked in one strand as a result of a partial digestion with the restriction enzyme in the presence of ethidium bromide. A short single stranded region is then exposed using the exonucleolytic activity of *E. coli* DNA polymerase polI) in the presence of a single deoxynucleoside triphosphate. This DNA is treated with sodium bisulphite, which deaminates C residues in single stranded DNA to give U. The gap can then be repaired with polI using all four deoxynucleoside triphosphates, resulting in the incorporation of an A residue to pair with the U. Replication *in vivo* then generates an A:T base pair. Base substitution can also be carried out by directing the incorporation of nucleotide analogues in the vicinity of restriction endonuclease cleavage sites [7]. A number of enzymes will cleave only one strand within a recognition site, under partial digestion conditions in the presence of ethidium bromide. A base analogue can then be introduced into the DNA by carrying out the nick-translation reaction catalysed by polI in the presence of limiting concentrations of the substrate triphosphates. This reaction involves the concerted action of the 5' → 3' exonuclease of *pol*I at the site of the nick, and the polymerising activity which replaces the excised mononucleotides with nucleoside triphosphates supplied to the reaction mixture. In this way dTMP residues can be replaced by hydroxy-dCMP. Nucleotide sequence analysis of plasmid DNA mutagenised in this way reveals that AT → GC transitions have taken place.

7.2 Expression systems

7.2.1 *Cloning in yeast*
One demonstration that yeast protoplasts could be transformed with DNA was the work of Hinnen *et al.* [8], who were able to select yeast cells which had taken up and expressed the yeast *leu*2 gene carried on Co1E1 (see Chapter 5). The transformants were analysed by the hybridisation technique of Southern (Chapter 6), and three classes identified in which sequences carried on the plasmid had been incorporated into yeast chromosomal DNA (Figure 7.2). Both the type I and type II events are proposed to arise from recombination events, in the first case through homology at or near the chromosomal *leu*2 locus, and in the second case at some other chromosomal site. The type I transformants are characterised by a larger *Hin*dIII restriction fragment (indicated in Figure 7.2), and by

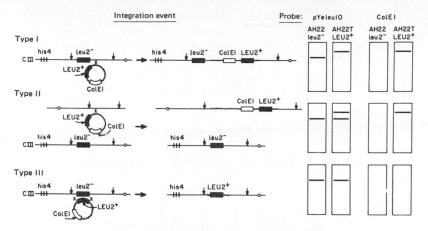

Fig. 7.2 The three types of yeast transformants. The arrows ($\downarrow$) on the physical maps represent *Hin d* III sites. Southern hybridisation patterns are shown diagrammatically on the right for *Hin*dIII cleaved DNA of the *leu2*⁻ recipient and its transformed derivative when probed with either the recombinant plasmid or ColEl vector sequences alone (from [8]).

genetic analysis which shows that the new *leu2*⁺ region is closely linked to the old *leu2*⁻ region. Similar restriction and genetic analysis of the type II transformants shows that in this case the integration of the plasmid is at a different chromosomal site, since the new *leu2*⁺ region is not linked to the old *leu2*⁻ region. In both these cases ColEl was shown to have integrated into the yeast chromosomal DNA where it segregates in a Mendelian manner. In the type III double-crossover integration event the vector sequences are not stably incorporated in the yeast chromosome.

The duplicated structure produced in the type I event is not stable. Struhl *et al.* [9] showed, in the case of a *his* transformant generated as described above, that after 15 generations approximately 1% of colonies were *his*⁻ segregants, and these cells had completely lost the transforming DNA by an intrachromosomal crossover. This observation has been used by Scherer and Davis [10] as the basis of a means for exchanging a chromosomal segment of yeast DNA for the homologous segment which contains an insertion or deletion introduced *in vitro*. The principle of this method is illustrated in Figure 7.3 for the yeast *his3* gene carried on a bacterial plasmid together with the yeast *ura3* gene. The *ura3* gene is used only as a selective marker and since the recipient yeast strain has an *ura3* deletion, homologous recombination occurs between the plasmid and the *his3* gene on the yeast chromosome. In this example, a 150 base pair deletion has been introduced into the *his3* gene on the plasmid by *Hin*dIII cleavage and religation. Following the selection of transformants, the selective pressure is removed and *ura*⁻ segregants are recovered. These have lost the *ura3*⁺ gene and linked bacterial sequences by recombination

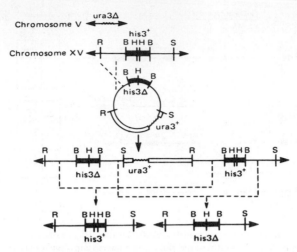

Fig. 7.3 Replacement of a yeast chromosomal segment with the *his*3 gene having a deletion. The deletion is carried on a plasmid which undergoes a type I integration event (See Fig. 7.2). Two classes of segregants arise from the duplicated structure as shown by the dashed line (from [10]).

between the pre-existing sequences at or around the *his*⁺ locus and the newly introduced homologous sequences. Two classes of segregants are identified, one of which is indistinguishable from the nontransformed strain, and the other in which the *his*3⁺ locus has been replaced by the *his*3 gene containing the 150 base pair deletion.

The transformation frequency of yeast cells by recombinant DNA molecules containing such markers as *his*3, *leu*2 and *ura*3 is about 10^{-7} transformants per viable cell. Segments of yeast DNA linked to the *trp*1 and *arg*4 genes have been demonstrated to transform at a much higher frequency (about 10^{-4}) [9, 11, 12]. In both cases the transforming molecule replicates autonomously within the cells. The transformants are unstable and the molecules are rapidly eliminated if selective pressure is taken away.

A second method of achieving efficient transformation of yeast cells uses a set of vectors derived from the endogenous 2 micron yeast plasmid [13]. The plasmid is present in many strains of yeast in 50–100 copies per cell. Its function is not known. Cloning vectors have been constructed which are *in vitro* recombinants between this plasmid and bacterial cloning vectors such as pMB9 or pBR322, and using such vectors it is possible to propagate cloned DNA segments in either *E. coli* or yeast cells. The bacterial gene for ampicillinase seems to be functionally expressed in yeast cells since, if such cells are plated with *E. coli* in the presence of ampicillin, the yeast colonies permit a halo of bacterial growth around themselves [14]. Initial studies on the expression of higher eukaryotic genes in yeast have been disappointing. The chromosomal gene for rabbit β globin has been introduced into yeast in a 2μ plasmid vector, and shown to produce globin-specific RNA which lack 20–40 residues at the 5′ end by comparison with mature rabbit β globin message, and which contain all of the small intervening sequence but terminate within the large intron [15].

7.2.2 SV40 as a cloning vehicle

Detailed knowledge of the molecular biology of the small DNA tumour virus [16, 17], simian virus 40, makes it a useful vector for introducing foreign DNA sequences into cultured mammalian cells. The virus can have two kinds of interaction with its host cell, depending upon the species. In permissive monkey cells one sees a productive infection cycle resulting in cell death and the release of progeny virions, whereas in non-permissive mouse or rat cells a small proportion of cells undergo heritable alterations in their growth properties, and are said to become transformed. Such transformed cells will produce tumours when injected into immunoincompetent animals. In non-permissive infections only the early viral genes are expressed, and in those cells which become stably transformed the viral DNA becomes integrated into the host cell chromosome. In the infection of permissive cells the virus is first transported to the nucleus where it is uncoated, after which stable early transcripts are produced which correspond to about one half of the genome (see Fig. 7.4). Complementation analysis of temperature-sensitive mutants indicates a single early complementation group, A. The product of this gene, the viral tumour antigen (T-ag), is required to initiate DNA replication which takes place bidirectionally from the single origin shown in Fig. 7.4. The early messenger RNAs code for a second tumour antigen,

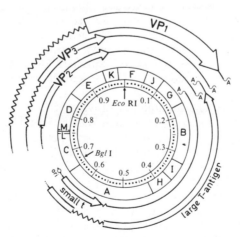

Fig. 7.4 The genome of SV40. The origin of replication is indicated by 'ori'. The outer circles show the locations and structures of stable cytoplasmic messages. Open blocks are coding sequences and the zig-zag lines represent sequences removed by splicing (from [18]; copyright © 1978 Macmillan Journals Limited).

71

a smaller related protein called 'small-t-antigen' (t-ag). The small t-ag seems to be essential for neither productive infection nor transformation. The two early genes contain non-coding intervening sequences which are removed from the mRNA by splicing. A larger intervening sequence is removed from the T-ag mRNA which is therefore the smaller message. Only some of these intervening sequences are removed from the small t-ag mRNA, and so the RNA molecule is larger. It encodes a smaller polypeptide, however, since it contains a termination codon in that part of its sequence not found in T-ag mRNA.

Late transcripts are found after the onset of viral DNA replication, and these encode the three capsid proteins VP1, VP2 and VP3. These mRNAs are transcripts of the opposite strand to the early mRNAs, and they each have 5' leader sequences which correspond to sequences on the genome which are well-removed from the coding sequence. The intervening sequences between the leaders and the body of the mRNA are again removed by splicing.

As with the case of bacteriophage λ, there is a physical limit to the amount of DNA which can be packaged into the virus capsid. Since the SV40 genome is small (5.2Kb) and since the three identified non-essential regions of the genome only amount to several hundred base pairs of DNA, then the use of non-defective viral vectors is limited. This has led to the use of defective viral vectors. The infection then has to be 'helped' by viral DNA which supplies the missing functions. One set of such defective viral vectors used mutants which have tandem repeats of the replication origin [19, 20]. These were cleaved to unit length and then ligated to foreign DNA. The infection of monkey cells with the DNA has to be helped with wild type viral DNA. This leads to the problem of the helper virus outgrowing the recombinant. In order to get around this problem, a helper virus can be used which carries a temperature sensitive mutation which is complemented by the segment of DNA within the vector [21]. The scheme for the use of such vectors is shown in Figure 7.5 for the specific case of SVGT-1, the BamHI/HpaII fragment of SV40 DNA which carries the early genes and replication origin. Recombinants between this DNA sequence and foreign DNA are introduced into monkey cells together with tsA mutant helper virus DNA. The infection of cells with the two virus types is then detected by the formation of plaques at 41°C. The plaques can be screened for the presence of specific sequences, using a procedure [22] analogous to those developed for the 'in situ' hybridisation of bacterial colonies or of plaques produced by phage λ (see Chapters 3 and 4).

A number of genes have been cloned into vectors such as SVGT-1 [21], and, although the foreign segments of DNA were transcribed, discrete functional RNA molecules were not seen. Mulligan et al. [23] argued that this was due to the deletion in these vectors of the viral DNA sequences required for processing late mRNA, the late leader sequence and the sequence in the body of the VP1 gene to which this is spliced. In order to leave these sites intact, they developed a

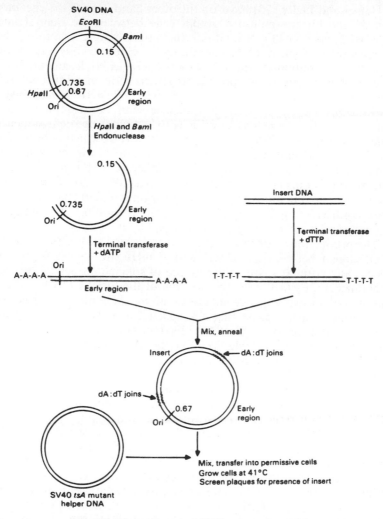

Fig. 7.5 Scheme for use of SV40 vectors (from [24]).

vector, SVGT-5, which lacks sequences between the *Hin*dIII site and *Bam*HI sites found internally within the sequence of the body of the VP1 gene. They used this vector to clone a segment of rabbit β globin cDNA containing the fMet-initiating codon 37 nucleotides from its *Hin*dIII terminus and the translation termination codon just proximal to its *Bgl*II terminus. When the recombinant was introduced into monkey cells together with helper SV40 *tsA* DNA, it was possible to detect discrete 1.8 and 1.0Kb RNA molecules which contained both SV40 leader sequences and β globin coding sequences. Furthermore, these cells were demonstrated to be synthesising a protein immunologically recognisable as rabbit β globin.

Hamer and Leder [25] have, on the other hand, cloned the chromosomal gene for the mouse β globin gene between the BamHI and HpaHII sites of SV40 DNA, and in this case they were able to see the production of stable mRNAs containing the globin coding sequences. The gene is contained within a 2.4Kb SacI/BamHI fragment. Both intervening sequences are present, together with an additional 430 residues to the 3' end. When ligated to the BamHI/HpaII-cleaved SV40 DNA this gives a linear molecule with non-complementary SacI and HpaII termini which cyclise in vivo, generating small sequence deletions at the site of recombination. The mRNA which is produced by the recombinant virus contains SV40 leader sequences and is approximately 300 bases longer than authentic globin mRNA. The intervening sequences are apparently correctly removed from the mouse gene by the monkey processing enzymes and, furthermore, the cells synthesise a polypeptide recognisable as globin both immunologically and by tryptic fingerprint analysis.

The production of unstable RNAs when other DNA segments have been cloned between the HpaII and BamHI sites of SV40 can be specifically correlated with the absence of splicing sequences. Recent experiments, in which both the position of the foreign DNA and the number of available splice junctions in the recombinant molecule have been varied, suggest that the splicing activity is necessary for the production of the stable RNA. Furthermore, any splice junction will suffice in this regard whether it be the SV40 specified sequence or sequences within the cloned gene [26].

7.2.3 Direct transformation of mammalian cells

As an alternative to the use of a viral DNA molecule as a vector DNA can be introduced directly into cultured mammalian cells along with some other DNA molecule which carries a selective marker. The thymidine kinase (tk) gene from herpes simplex virus or from mammalian DNA can be transferred into cells deficient for this enzyme (tk$^-$), and transformants can be selected by their acquisition of the ability to grow in media containing hypoxanthine, aminopterin and thymidine. The transformants arise from a sub-population of cells which are competent and also take up other DNA molecules not physically linked to the DNA carrying the thymidine kinase gene [27]. In this way the genomic sequences for the rabbit β globin gene carried on phage λ DNA have been introduced into tk$^-$ mouse L cells. The globin gene and its flanking sequences are contained within this phage in a 4.7Kb KpnI fragment which can be detected in the chromosomal DNA of the transformants. Analysis of the rabbit β globin transcripts from one of these clones (using the techniques of Alwine et al. and Berk and Sharp described in Chapter 6) indicates that the intervening sequences are correctly removed from the RNA, but that the 5' end of the RNA molecule lacks some fifty residues found in mature rabbit β globin mRNA [28]. Similar experiments have also been carried out with long concatamers of the rabbit β globin chromosomal gene covalently linked to pBR322 carrying the herpes tk gene. Cells transformed with these concatamers synthesise β globin mRNA which is indistinguishable from the natural mRNA [29]. The human chromosomal β globin gene sub-cloned from a bacteriophage λ vector into pBR322 has also been introduced into tk$^-$ L cells. In this case restriction analysis of the DNA from the transformed cells is consistent with the continued existence of the β globin plasmid as an autonomously replicating plasmid within the cells with some integration into high molecular weight DNA [30]. This DNA segment may therefore contain a functional mammalian DNA replication origin and so may find future application as a vector for introducing foreign DNA into mammalian cells.

7.2.4 Microinjection of cloned DNAs into Xenopus laevis oocytes
When SV40 DNA is microinjected into the nuclei of *Xenopus laevis* oocytes it directs the synthesis of the SV40 virion proteins VP1 and VP3 [31]. The DNA must therefore be not only correctly transcribed, but also the transcript must be correctly spliced, transported to the cytoplasm and translated. Ribosomal genes are also transcriptionally active when injected into the oocyte nucleus. An analysis using the electron microscope shows that, following injection, the plasmid DNA is assembled into chromatin of which some 10% is seen in association with nascent chains of ribonucleoprotein fibrils. The high packing density of the growing transcripts and the smooth appearance of the non transcribed spacer resemble the appearance of endogenous transcription units [32] (Figure 7.6). Genes transcribed by RNA polymerase III are expressed very efficiently as shown for injected 5S genes [33] and tRNA genes from *Xenopus* and yeast [34, 35]. The latter study is particularly interesting since it examined the transcription of the cloned yeast gene for tRNA^{tyr} which contains a 14 base pair intervening sequence. The gene is transcribed in the oocyte to give an RNA with a 5′ leader extension and containing the intervening sequence, both of which are correctly removed. Furthermore, several post-transcriptional modifications are correctly carried out, including addition of CCA to the 3′ terminus of the molecule. Not all of these steps are carried out equally efficiently, and the removal of the intervening sequence occurs in only about 30% of the molecules. The experiment does, however, illustrate that the enzymes necessary for tRNA processing have been well conserved through the period of evolution in which yeast and frogs have diverged. It should then be no surprise to learn that sequences coding for several tRNAs have been recognised on a cloned 3Kb segment of *X. laevis* DNA, and that some of these genes contain intervening sequences at a position analogous to that in yeast tRNA genes.

The *Xenopus* oocyte nucleus seems to be particularly rich in RNA polymerase III, and considerable success has been achieved with extracts from these cells in the cell-free transcription of tRNA genes and 5S DNA [36]. The transcription of the cloned somatic 5S rRNA gene

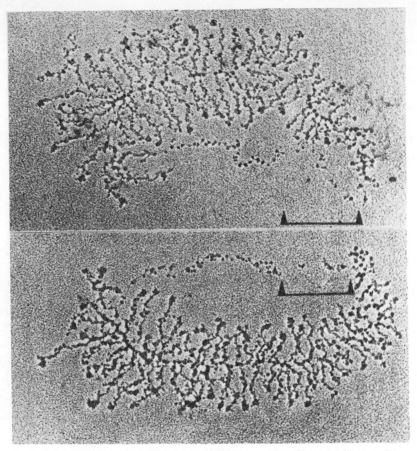

Fig. 7.6 Transcription patterns of cloned rDNA injected into *Xenopus* oocyte nuclei. The spacer sequences have an unblocked appearance (arrows) unlike the plasmid vector DNA (from [32]; copyright 1978 Macmillan Journals Limited).

from *X. borealis* has been examined in such a system. A number of cloned templates were tested in which enzymatically generated deletions extended into the gene from either the 5' or 3' side. This has surprisingly revealed a region *within* that part of the gene coding for residues 55–83, which is required to direct RNA polymerase to its initiation site. If sequences to the 5' side of this region are deleted then transcription still occurs to produce an RNA molecule of approximately the correct size, but with incorrect 5' sequences [37].

Recently systems have been developed in which RNA polymerase II will correctly initiate transcription but in which termination is poor [38, 39]. Specific 'run-off' transcripts can be made, however, when a restriction fragment specifically cleaved within a gene is used as a template. Wasylyk *et al.* have used such a system to analyse the effect of deleting sequences from the cloned chicken conalbumin gene [39]. They find that *in vitro* transcription is entirely eliminated by deleting the sequences from −44 to −8 residues upstream from the 5′ end of the transcription unit, a region which includes the TATAAA box.

As such systems continue to be developed, then we will surely see the concerted application of *in vitro* recombination and mutagenesis in a new chapter in the study of eukaryotic gene expression.

References

[1] 'The Hogness Box'. Goldberg, M. and Hogness, D. S. Personal Communication.

[2] Breathnach, R., Benoist, C., O'Hara, K., Gannon, F. and Chambon, P. *Proc. natn Acad. Sci.* **75**, 4853.

[3] Sakano, H., Huppi, K., Heinrich, G. and Tonegawa, S. (1979), *Nature* **280**, 288.

[4] Carbon, J., Shenk, T. E. and Berg, P. (1975) *Proc. natn Acad. Sci. USA.* **72**, 1392.

[5] Covey, E., Richardson, D. and Carbon, J. (1976) *Mol. gen. Genet.* **145**, 155.

[6] Shortle, D. and Nathans, D. (1978) *Proc. natn Acad. Sci. USA* **75**, 2170.

[7] Muller, W., Weber, H., Meyer, F. and Weissmann, C. (1978) *J. mol. Biol.* **112**, 343.

[8] Hinnen, A., Hicks, J. B. and Fink, G. R. (1978) *Proc. natn Acad. Sci. USA* **75**, 1929.

[9] Struhl, K., Stinchcomb, D. T., Scherer, S. and Davis, R. W. (1979) *Proc. natn Acad. Sci. USA.* **76**, 1035.

[10] Scherer, S. and Davis, R. W. (1979) *Proc. natn Acad. Sci USA* **76**, 4951.

[11] Hsiao, C-L, and Carbon, J. (1979) *Proc. natn Acad. Sci. USA* **76**, 3829.

[12] Stinchcomb, D. T., Struhl, K. and Davis, R. W. (1979) *Nature* **282**, 39.

[13] Beggs, J. D. (1978) *Nature* **275**, 104.

[14] Hollenberg C. (1979) personal communication.

[15] Beggs, J. D., Van der Berg, J., Van Ooyen, A. and Weissmann, C. (1980) *Nature* **283**, 835.

[16] A useful short review by Rigby, P. W. J. (1979) *Nature* **282**, 781.

[17] A longer review by Fareed, G. C. and Davoli, D. (1977) *Ann. Rev. Biochem* **46**, 471.

[18] Fiers, W., Contreras, R., Haegeman, G., Rogiers, R., Van de Voorde, A., Van Heuverswyn, H., Van Herreweghe, J., Volckaert,G. and Ysebaert, M. (1978), *Nature*, **273**, 114.

[19] Nassbaum, A. L., Davoli, E., Ganem, D. and Fareed, G. C. (1976) *Proc. natn. Acad. USA.* **73**, 1068.

[20] Ganem, D., Nassbaum, A. L., Davoli, D. and Fareed, G. C. (1976) *Cell* **7**, 349.

[21] Goff, S. P. and Berg, P. (1976) *Cell* **9**, 695.

[22] Villareal, L. P. and Berg, P. (1977) *Science* **196**, 183.

[23] Mulligan, R. C., Howard, B. H. and Berg, P. (1979). *Nature* **277**, 108.

[24] Rigby, P. W. J. (1979) *Biochem. Soc. Symp.* **44**, 89.

[25] Hamer, D. H. and Leder, P. (1979) *Nature* **281**, 35.

[26] Hamer, D. H. and Leder, P. (1979) *Cell* **18**, 1299.

[27] Wigler, M., Sweet, R., Sim, G. K., Wold, B., Pellicer, A., Lacy, E., Maniatis, T., Silverstein, S. and Axel, R. (1979) *Cell* **16**, 777.

[28] Wold, B., Wigler, M., Lacy, E., Maniatis, T., Silverstein, S. and Axel, R. (1979) *Proc. natn. Acad. Sci. USA* **76**, 5684.

[29] Mantei, N., Boll, W. and Weissmann, C. (1979) *Nature* **281**, 40.
[30] Huttner, K. M., Scangos, G. A. and Ruddl, F. H. (1979) *Proc. natn Acad. Sci. USA*. **76**, 5820.
[31] DeRobertis, E. M. and Mertz, J. E. (1977) *Cell* **12**, 175–182.
[32] Trandelenberg, M. F. and Gurdon, J. B. (1978) *Nature* **276**, 292–294.
[33] Brown, D. D. and Gurdon, J. B. (1977) *Proc. natn Acad. Sci. USA*. **74**, 2064–2068.
[34] Kressman, A., Clarkson, S. G., Pirotta, V. and Birnsteil, M. L. (1978) *Proc. natn Acad. Sci. USA*. **75**, 1176–1180.
[35] De Robertis, E. M. and Olson, M. J. (1979) *Nature* **278**, 137–143.
[36] Birkenmeier, E. H., Brown, D. D. and Jordan, E. (1978) *Cell* **15**, 1077–1086.
[37] Sakonju, S., Bogenhagen, D. F. and Brown, D. D. (1980) *Cell* **19**, 13.
[38] Weil, P. A., Luse, D. S., Segall, J. and Roeder R. G. (1979) *Cell* **18**, 469.
[39] Waslyk, B., Kedinger, C., Corden, J., Brison, D. and Chambon, P. (1980) *Nature* **285**, 367.

Index